a father's Sexting teen

The Brian Hunt Story

Cape Cod, Massachusetts

by Annie Winston

Published by Tri-Net Publishing, LLC™ | Irvine, California

Cover design | Buz Phelan

All rights reserved. No part of this book may be reproduced in any manner whatsoever without written permission except in the case of brief quotations used in critical articles and reviews. The Brian Hunt story is based on a true story.

© 2010 Annie Winston and Brian Hunt

ISBN 9781456334444

"All I ever cared about was protecting my thirteen-year old-son from the moment I found out that he was facing a prison sentence if convicted of the felony charges of being a child pornographer and sex offender for passing on from his cell phone to another cell phone, a photo of a bare breasted female student. I'm going to fight this right down to the end and if necessary I'll bring the bigger guys in (the media) because what is happening is not right! I will never give up no matter the cost to my reputation or finances. My son's life won't be ruined."

— **Brian Hunt's statement to a Cape Cod Time's Reporter**

One reporter at "ABC News" said they wanted an interview with Brian Hunt because he was the first parent of a student caught up in a sexting controversy, nationwide, to stand up and fight on behalf of the kids.

— **Falmouth Enterprise | February 13, 2009**

"You made a comment that you think there ought to be warnings in cell phones when you give them to a kid that explains this and that they have to let their parents know that they understand that. I think that's a really, really good idea, and I'm glad that you brought it up."

— **Dr. Phil to Brian Hunt | May 9, 2009**

"What an important book for such a serious issue in these times. The teen sexting problem is like having a fire spewing dragon in your backyard! You simply can't ignore it.

If you do, it just might 'scorch' your teen and family!

> — **Jim Burns, International Author/Speaker on Family issues and President of Homeword.**

"A Father's Sexting Teen' is a compelling and engaging narrative of the continually evolving challenges schools and families face at the crossroads of technology and children. How do we protect our children when they and their peers know more about the latest communications media than we do? A tough problem, dealt with compassionately and maturely in this story. Bravo!"

> — **Glenn Noreen, MBA, Harvard Business School and Ingenium, Executive Director | Barack Obama Charter School | Compton CA**

"This book is a must read for all parents of teenagers! Brian is an average American father who has risked his own reputation to warn parents and teens about the potential legal disasters and severe consequences that can result from sexting. The majority of people do not understand the gravity of this issue, but what has happened to the Hunt family can easily happen to any one of us, as well."

> — **Dr. Charles Sophy, an advisory council member of LG Text Ed, a program of LG Mobile Phones and nationally recognized family psychiatrist.**

[SEXT • ING] SEX (passing on illicit photos or videos via cell phones) + **TEXTING** (intimate messages sent over cell phones) = **SEXTING**

One in every five teens has sent intimate text messages or posted revealing photographs of themselves, with 71% of teen girls and 69% of teen boys stating that they sent these pictures to their boyfriend or girlfriend. 44% of both teen girls and boys say it is common for such text messages and photographs to be shown or shared with others, while 36% of teen girls and 39% of teen boys said that it was common for nude or semi-nude photographs to be shared with people other than the intended receiver. National Campaign to Prevent Teen and Unplanned Pregnancy conducted a survey that deals expressly with the issues raised by recent events. The survey results are available in PDF format on their website:

www.theNationalCampaign.org/SexTech/PDF/SexTech

When my dear fiends Elaine and Allen Niemi first called and asked if I would be interested in writing a book about teen sexting, my first was "you've got to be kidding! I write children's books about heroes "You've got to be kidding! But after I spoke with Brian and hear what he did to stand up for his son I thought this guy is a hero He's someone who is willing to fight for what's right when it isn't easy. I couldn't wait to start writing his story. It's my hope you'll enjoy the book as much as I did in writing it. God Bless You!

— **Annie Winston**

Acknowledgements

A big thank you goes to Brian Hunt and his family, my friends and business associates and most importantly my three teens who believed in this project and were willing to offer support, prayers and encouragement by doing whatever it took to help make it happen. You are appreciated!

I'm Brian Hunt and I'm not much of a writer, but I wanted to tell my story and that's why I asked Annie Winston to help me out. You might wonder what a book is all about that's called "A Father's Sexting Teen." If you don't know what a "sexting teen" is, it's a teen who sends sexually explicit photographs of either themselves or of someone else via a cell phone, instant message, text or email.

On February 9, 2009, I was very unhappy to find out my thirteen-year-old son became a sexting teen after he pushed the send button on his cell phone and forwarded a photo of a bare breasted female student. His action brought him to the Falmouth Police where he and five other Lawrence Junior High School students faced felony charges for trafficking child pornography. In the days that followed, my son Ben and myself became the focus of a national media frenzy on the subject of teen sexting. I kept a few scratch notes here and there with the thought of writing about what I went through so that the lessons I learned could be shared with other parents and teens because I don't want anyone else to go through what happened to me.

— **Brian Hunt**

Dedication

This book is written in memory of those teens who took their lives because the consequences of their sexting became too great to bear. Dedicated to all teens who have the courage to say

No to sexting and Yes to purity.

Table of Contents

After it Happened

Blog Entry 1 of 10
Reeling in shock............................11

Blog Entry 2 of 10
A Father and Son Chat.....................15

Blog Entry 3 of 10
Scare You Out of Your Mind...............20

Blog Entry 4 of 10
On the Ten o'clock News...................24

Blog Entry 5 of 10
Popcorn Ceiling and God.......................27

Blog Entry 6 of 10
Lights, Camera, Action 30

Blog Entry 7 of 10
Missed Court Appearance......................34

Blog Entry 8 of 10
Agreeable & Disagreeable Opinions.....40

Blog Entry 9 of 10
Reflections..42

Blog Entry 10 of 10
Down and up...45

Media Frenzy

Blog Entry 11 of 20
A Father's Heart 48

Blog Entry 12 of 20
Mouthing Off 49

Blog Entry 13 of 20
Coffee Musings 54

Blog Entry 14 of 20
To New York City 56

Blog Entry 15 of 20
A Star I Am Not.........................58

Blog Entry 16 of 20
Walking, Walking, Walking.......62

Blog Entry 17 of 20
Not Something to Brag About.....63

Blog Entry 18 of 20
No Worries…..…………........66

Blog Entry 19 of 20
Small Town 68

Blog Entry 20 of 20
Nervous 71

Everybody's Talking

Blog Entry 21 of 30
Can't sleep.................................72

Blog Entry 22 of 30
Like a Baby 74

Blog Entry 23 of 30
Hurry up and Wait 75

Blog Entry 24 of 30
Proud 79

Blog Entry 25 of 30
Vultures 80
Blog Entry 26 of 30
"I Wish I Hadn't"................ 87
Blog Entry 27 of 30
Anticipation Makes Me Crazy. 91
Blog Entry 28 of 30
Round of Applause...................92
Blog Entry 29 of 30
What If 97
Blog Entry 30 of 30
It's a Big Deal..........................100

Epilog   103

Life Building Insight   107

What parents can do to prevent teen sexting

What parents can do if your teen is caught sexting.

Eight things to know if you're a sexting teen

Real Voices for Pure Fun

Aaron Craver.	118
Kandee Johnson.	121
Jim Burns.	125
Randy Largent .	127
Annie Winston.	129
Sponsored Links	131

Blog Entry 1 of 10 – After It Happened
Reeling in Shock
Monday | February 9, 2009 | 2:03 p.m.

A storm swirls outside. An IV is stuck in my arm, and a white bag of liquid food drips into my veins through a tube that is wrapped up and around my chest.

Diverticulitis surgery is not fun but it's way better than the pain. A nurse comes by and asks me if there is anything she can get for me. I let her know that I'm fine though I would really like a cheeseburger with onion rings.

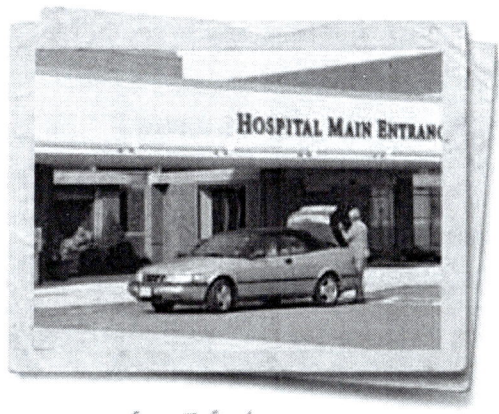

where I had surgery

The phone rings, maybe it's my wife; she usually calls at this time. I answer it and discover it's not her; instead a stern male voice is on the other end.

"Mr. Hunt this is the Vice Principal of your son's school

I need to talk to you about an important matter. Do you have a moment?" I think, Of course I have a moment, that's all I have lying around in a hospital bed. "So what's going on?"

"Mr. Hunt, your son is in my office for sexting." "Sexting?

What's that?" "Sexting is sending explicit sexual material from one cell phone to another. Your son, Ben, did this he has already been called in to my office and the Falmouth Police Department is involved.

"Sexually explicit material?" I'm reeling in shock. I can hardly believe what I'm hearing.

"Of what?" "A female student showing her breasts."

"Are you listening Mr. Hunt, we had to turn the case over to the Falmouth Police Department because your son and five other boys face serious trouble with the law."

"Serious trouble with the law?" What could that mean? Like a wildly speeding car on a curvy road, my mind raced and veered in all directions. Would my son be charged as a sex offender or, even worse, a child pornographer? The horrible thought hammered my mind like hailstones hitting a windshield. This I know, my son is not some thirty-year-old guy with a warped conscience preying on children for perverted pleasure. My son is a boy; he's only a thirteen-year-old kid and in many ways still a child. He had no idea that by clicking the send button on his cell phone that he could be charged with a federal offense and have it on his record until he was forty-four years old. A record like this will destroy his life from getting into a college to working as a dish washer.

Here's what I think went down on the day the photo arrived in his cell phone: Ben sat at his school desk gazing off, bored, as a cow might stare at a gate. Like a neon light flashing, his moment brightened when a text popped into his inbox with a photograph as an attachment. He opened it and experienced a rush of healthy teenage male hormones when his eyes feasted on a grand display of female beauty. Eve was in the garden and now technology had brought her to his pocket-sized cell phone. Eager to share the new wonder of the world, my son clicked the send button and passed the photo on to a school buddy who then passed it on to their buddies; and so over the cyber cell phone world the photo of the bare breasted beauty traveled faster than the speed of light. Before long, the photo was probably seen by just about every male student at Lawrence Junior High School, so

Lawrence Junior High

why was my son, and five other boys, the only ones brought to account for an act of stupid impulse. This isn't right.

"Did you hear me, Mr. Hunt? The Falmouth Police have been called as well as a state police trooper and a thorough investigation is beginning. We will get to the truth. Your son's phone has been confiscated." Disbelief that this was happening almost rolled me out of my hospital bed and onto the floor. My fight to protect my thirteen year old son from the charge of being a child pornographer and sex offender began.

Blog Entry 2 of 10 – After It Happened
A Father and Son Chat Tuesday | February 10, 2009 | 11:00 a.m.

It feels good to be home. Hospital visits are not my idea of getting away. My wife picked me up around 8 a.m. this morning and told me Ben stayed home from school today. I was glad he did as I wanted to look him straight in the eye and talk with him. Hearing his side of the story was important to me. Ben knew the "father son chat" was coming. He wasn't disagreeable to the idea. I'm very thankful that we've always had the kind of relationship where he feels safe opening up about what's going on with him. When he tells me something, I don't doubt his word.

Ben started out his story by saying the girl in the photo only showed one of her breasts and it wasn't a double whammy. I believe my son. I wonder why the Vice Principal didn't mention this? Ben said when he was called into the office, he had no idea why he was going in; he thought maybe it was to pick up some extra money for food as he had been sitting in Science class right before lunch. He told me before he heard his name called that he was feeling so bored he thought picking onions in a dry, hot field would have been better than listening to his teacher talk about protoplasts and zygotes.

Ben said the moment he received the flashing red light on his phone, telling him a text had arrived, he quickly opened it and boredom left him like a stampede of horses out the gate. The girl's glory met his stare and then Ben on not-so-smart impulse quickly pushed the send button to forward her picture to a school buddy. Ben told me he thought it was "just a picture" and he'd play a "little

my son Ben

joke" on his friends by sending it on. I stopped him. "Ben it's never a joke to pass on a bare breasted photo of one of your female classmates."

"Dad, you got to know this, I really wasn't thinking. It was only about ten minutes after I passed on the girl's picture when I heard my name called over the loud speaker telling me to come to the office. When I got there, the school secretary took me in to see a school administrator who sat in a large

A Father's Sexting Teen

overstuffed chair waiting behind a desk, looking like a judge in a courtroom who had already pronounced the criminal guilty."

Ben told me that the administrator also reminded him of a red clawed lobster about to catch his prey.

After he got to the Vice Principals office, the door was shut and then it was only he and the "lobster." One of the first things Ben was asked was whether or not he had the picture on his cell phone of the girl student. Ben said. "I'm not going to lie I do have the photo on my phone." "Then I want to see it." Then Ben was told to come up to his

Ben's cell phone

desk and flip open his cell phone. After the girl's glory was seen, Ben said the phone was "snagged" from him like someone reeling in a trout, fast and furious, and then it was put into the desk drawer." Ben thought it was not right for the Vice Principal to look at the girl's picture. I agree with Ben. A female secretary could have been called in to confirm it was there.

Annie Winston

What fries my egg is that my son's phone was taken from him without my knowledge or permission. I remember from my high school Government and Economics classes something about the Fourth Amendment to the Constitution which protects a citizen's rights from an unreasonable and unlawful search and seizure.

Once the vice principal had the cell phone, he left the room but not before telling Ben that it was going to be "forensically analyzed" (whatever that means) by the police and he had to wait and not leave the office. Ben said he wished he was back in his teacher's classroom listening to her boring lectures on zygotes and protoplasts. He would take almost any punishment over feeling panicked and wondering what was going to happen.

Lunch was brought to Ben, but the greasy cafeteria ham and cheese sandwich made him feel sick making his already queasy stomach even worse. The more Ben thought about what he might be facing the queasier he felt. He knew he wasn't going on a trip to Disneyland where he'd be shaking Mickey Mouse's hand. I can say my father and son chat went well. Ben felt understood. He was willing to take responsibility for his wrong but he didn't think being charged as a sex offender and going to jail was fair or the right punishment. I told him. "Ben it's not fair. Laws that were written for adult sex offenders should not be applied to thirteen year-old kids. I'll fight for you and the other boys and do everything I can to keep you out of jail!"

"Thanks dad!" Ben hugged me.

I told him. "Son, respecting another's private parts, especially a young girl, is always the best thing to do. Passing on the photo was a violation of her privacy, even if she didn't think so because she allowed the picture to be taken. Never forget this important life rule. Do to others what you would want them to do to you. Always ask yourself this question. If I do this, will mom and dad be proud of me? If not, then don't do it!"

Ben said he was sorry. I wouldn't be a good or loving father if I didn't discipline him. I grounded him for a week and said. "Don't forget sorry is made real when you try your best not to do again what you're sorry you did."

"Thanks dad, I love you!"

"I love you too!"

"Ben, do you remember Spiderman's line at the end of the Spiderman 3 Movie?" "Yeah, I can quote what Peter Parker says perfectly." "Go ahead!" "It's the choices that make us who we are and we can always choose to do what's right."

We high-fived on that one!

Blog Entry 3 of 10 – After It Happened
"Scare You Out of Your Mind" Tuesday
February 10, 2009 | 4:11 p.m.

Since Ben's cell phone is hanging out in the Vice Principal's desk drawer or somewhere in a police station lab being "forensically analyzed," I need to buy him another one for emergencies.

Once the Lawrence Junior High student's Sexting Teen story breaks in the news, I have a strong hunch the gossipers and letter writers to the newspapers will be clucking their tongues saying. "What an awful parent Brian Hunt is! He shouldn't reward his son for bad behavior by going out and buying him another cell phone; he's just going to send a nasty picture over it again!"

Gossips are like gophers, lying around just itching in the dirt for something to say about someone they don't like. Whatever the gossip gophers say now or will say, does not matter because I don't care. Protecting my son is what's important. I want Ben to have a phone so he can call me if he's in trouble. Ben's safety comes first.

The pain meds my doctor put me on are blurring my brain. I can't think straight. I should try and rest but I need to do some venting. I wanted to go down to the police station last night with Ben and my wife, Kim, but I couldn't. Ben's Uncle went instead and Kim told me all about it when she got home. She said it was "a scare you out of your mind" meeting.

An officer of the law introduced himself as the one in charge of the investigation. Kim told me that he was quick to point out that he was highly trained in internet crimes against children. He said "Your boys have a good chance of being charged and prosecuted with two felonies, possessing and distributing child pornography." He also made it clear that the boys ran a high risk of also being charged as sex offenders and it could stay on their record for forty years.

I got angry thinking about what Kim was telling me. I told her. "Come on; give the kids a break! They did what thirteen year-old kids do who have thirteen-year-old minds. Don't make them criminals because they use poor judgment!" As I'm writing this, even now I feel my father polar bear protection instincts surging in my veins, like a rushing stream through a mountain pass. My resolve to fight is an iron rope. I will do everything I can to make sure that Ben and the other students won't be victims of unfair laws that don't apply to teens who get caught sexting.

Stupid impulse is not predatory intent.

Kim also said that the officer told the parents that the cell phones were going to be "forensically analyzed" (there's those words again) and that their phone records had already been subpoenaed but, not to worry, because the boys were given a "break" from the law. They could have been arrested and arraigned on the charges immediately instead of

the "scare you out of your mind" meeting place

attending juvenile court first. The boys were going to be summonsed in Falmouth District Court and face a juvenile clerk magistrate hearing rather than an arraignment.

The officer is certainly doing his best to give the impression to the parents and community that the sexting incident is a very serious matter but no one should be worried because from the beginning of the investigation as he says "the goal has always been to protect the kids."

"Protect the kids?" How is it protecting the kids to stampede, intimidate and blow them out of the water by misapplying strict child pornography laws which were never written for teens who sext?

I understand that teens who sext are violating laws written to protect children from adult pornographers, and that the school has a legal responsibility to notify the police but appropriate punishment, PLEEZE! Taking a teen to jail and slapping him with a sex offender label for an unthinking act of not understanding the legal consequences of sending a nude or semi-nude photo of another teen is beyond over the top, how about one thousand miles high over the top?

I'm exhausted, still feeling raw after the surgery. It's not too often that one has twelve inches of their large intestine removed. I'm going to take a nap before I head out to Ben's basketball game.

Before I get a little shut eye, I need to vent about some of the news reporting. It was reported in this morning's paper that the boy's cell phones were taken from them on January 15, over three weeks ago. If that's true then Chicken Little was right when he ran around screaming at the top of his lungs that the sky was falling. Yesterday, Ben had his cell phone with him when he was called to meet with the Vice Principal. It was only then when Ben's cell phone was taken and tucked away in the Vice Principal's desk drawer. After I read the article, I couldn't help remembering what mom always told me growing up. She used to say, "Brian, don't believe everything you read in the newspaper." The older and wiser I get, I can't agree with her more.

**Blog Entry 4 of 10 – After It Happened
On the Ten o'clock News
Wednesday | February 11, 2009 | 8:39 a.m.**

Last night I was on the ten o'clock news. Felt strange seeing myself on TV. I've heard some say that you always look five to ten pounds heavier. I think I looked fifty pounds heavier. After I got a little shut eye yesterday afternoon, I left the house around six for Ben's basketball game. My plan was to come home afterwards and spend a quiet evening with Kim watching TV and then go to bed. Little did I realize, I would be stepping on a twisty looping roller coaster and going for a ride that I'd be on for the next several months. It all began as I stood outside the gymnasium waiting for Ben to come out of the locker room. I overheard a conversation between two students and a news reporter from the local TV Station, FOX 25.

"Ben Hunt was one of the sexting guys called into the Vice Principal's office. His dad is standing right over there." At that moment, I felt shame and embarrassment. I wished the whole thing had never happened. I wanted to dash and disappear and not deal with the reporter. The only thought running through my brain was

"I am **not** the father of a sexting teen.

I am **not** the father of a sexting teen

I am **not** the father of a sexting teen."

I flashed back to Ben's baby and childhood school photos on our living room wall. I saw Ben's kindergarten picture where he was missing his front tooth and then his fourth grade picture where he looked like a pickle in a vinegar jar. My sweet baby, pickled face fourth grader, was now facing felony charges. I never want to see a booking mug on our family wall. Still wanting to bolt, I turned around to escape, but the reporter yelled my name.

"Mr. Hunt, can I talk to you? I'm a news reporter from Channel 25 FOX news. If you don't mind, I would like to ask you a few questions. Is it true that you're the father of a sexting teen?"

"Yes, I am the father of a sexting teen." The words were out and so were my cowardly lions of shame, and embarrassment. My courage and determination to fight for Ben and the boys returned. My iron rope resolve won't be broken. Cowardice and shame won't weaken it. My story will be told to the world, or at least to FOX 25. My son's life won't be ruined and this tsunami wave won't drag me down and leave me choking on sand.

I will persevere and get through it and I will do whatever it takes to get my son free and out of the "curse" of the law, even if it meant talking to reporters and other media journalists. Running the risk of being misquoted, criticized and looking larger than I want to on television is a small price to pay for getting my son and the other boys off the penalty of "death." I will pursue truth and what is right until victory is in my hand.

The reporter asked if I would be willing to be interviewed at my home and if he could follow me back to my house. "No problem." Before long I had a news truck in my front driveway and a microphone in my face. I told them, "My son is a good kid as well as the others. They were taught a lesson and they know to never do it again."

At the time I was telling my story in my driveway to FOX News, back at Lawrence Junior High, a school committee meeting was going on. A friend called to tell me that the Boston-based Channel 7 and Hyannis Radio station WQRC were milling around outside the school campus looking for anyone to talk to. They also went inside the building where the meeting was happening. The School Superintendent was trying hard to explain everything going on with the sexting issue. A student's father stood up and insisted on getting clear answers. "How come all the students, administrators and Falmouth Police Department knew about the problems with sexting at the school before the parents?"

My friend said that the Superintendent would be sending out a pre-recorded phone message about the sexting issue and it would go out to every parent at Lawrence Junior High. Some parents in the audience were upset, saying "Why weren't we told about the sexting issue when it first came up three weeks ago?" No comment.

Blog Entry 5 of 10 – After It Happened
Popcorn Ceiling and God Wednesday
February 11, 2009 | 11:30 a.m.

The phone rang around 11 o'clock last night, right after I finished recording my interview on my VCR I knew who was calling. I can't stand pre-recorded messages, especially this one. I let the answering machine take it. I went to bed worried and stared at the popcorn ceiling wondering if talking to God about what was going on would make a difference. Would my prayers go higher than the ceiling? I'm not a religious guy but I went ahead and prayed and got very real with God, just in case it worked. I told Him how frustrated I was and I would really like Him to get Ben and the boys out of danger.

I've often thought about which God is the true one. The world and all its belief systems have a whole bunch of different ideas about God. I heard someone say "what you spend the most time thinking about and doing and spending your money on, then that's your god." From the time I was a small boy, I was taught the Catechism at St. Patrick's Church. The Catechism told me who the true God was, the Creator of everything and everyone. The idea was simple and it made sense. I'm just not sure what it means. I suppose it means treating others like the way you want to be treated, and doing your best to make God important in your life. I suppose I should be going to church more than only on Easter and Christmas. *...where I learned the Catechism*

I can't forget the memories I have of myself in church as a young boy wiggling in my seat sitting next to my six brothers and sisters, father and mother. At the time, I thought the Priest leading the service had a steeple stuck in his throat because he was so solemn and tight when he led us in saying out loud the catechism. "I believe in God the Father Almighty, Creator of Heaven and Earth...

I know my mother believes with her whole heart, and she also believes in the power of prayer. She says that prayer is always answered just not in the ways you think. I hope she's right.

I still couldn't sleep after talking to God but I felt better. I glanced at the night stand clock. It was three o'clock in the morning. I decided to get up. I went to the kitchen and listened to my phone messages. I was right. The eleven o'clock caller last night was a computer leaving the School Superintendent's pre-recorded message. He said parents shouldn't worry about the sexting issue. School authorities and police had it under control. I don't agree.

"Shouldn't worry?" I could hardly sleep last night. If anyone should ask what I thought about the whole matter, I'd tell them! "Too little, too late." Don't wait three weeks before the parents are notified about an investigation that involves their student and don't let the students and police know about "the problem" before the parents do. Better go, the phone's ringing. I hope someone wants to rent my limo. I haven't had a booking in awhile. Even though I'm not feeling totally back to normal, I'll take the job because I need the cash to feed my family and pay bills.

Blog Entry 6 of 10 – After It Happened
Lights, Camera, Action
Thursday | February 12, 2009 | 12:15 p.m.

I'm hungry. I want to eat a bologna sandwich. I haven't had time to make one. The phone's been ringing off the hook and the calls haven't been from folks needing to book a limo. They've been from reporters all over the country. What's sparking their interest? Some of them must have seen my FOX 25 News interview and how young teens are being treated like adult sex offenders when they unknowingly violate child pornography laws. They're also seeing one of the fathers, a big guy, willing to stick his neck out for his teenage son and the other boys. Not only was the "Cape Cod Times," "Falmouth Enterprise," "Boston Globe" and "Falmouth Bulletin" calling but "Geraldo," "Inside Edition," "The O'Reilly Factor," "The New York Times" and "ABC News" were ringing my phone.

"ABC News" was in my living room this morning, moving my coffee table and easy chair. A film crew was sent over and before long it was "lights, camera, and action." I told my story, this time it was easier than my first FOX 25 interview.

"Inside Edition" also called wanting Ben and me to go on their show with one of the other dads and his son who also got caught sexting. I haven't given one of the producers, my final answer yet. If I go on, I'll tell my story not only to get my son and the rest of the boys off the chopping block but to help other families and teens who may one day find themselves where I am. Those who criticize me for going public and say I'm only flapping my gums because I'm trying to make a name for myself. To those I say, "you've got to be kidding!" Not at this price! The price of feeling embarrassed, shamed and being thought of by some as a "failure of a dad" because my son "sexted." While dropping Ben off at school this morning, I was greeted by large media trucks with huge almost space ship sized satellite dishes parked near the school, town lots and in Falmouth neighborhoods. I'm glad the story is getting out and the media is coming. It's this national exposure that will put pressure on the police department and courts to lighten up on the kids.

Ben came home from school yesterday and told me that the officer who was heading up the investigation spoke to the students at a special assembly about the dangers of sexting. Ben said that he came down very hard on the issue, just like he did the other night at the police station when he

> **Media 'Firestorm' Over Phone Nudity Case**

talked to the kids and the parents. Frightening the kids with the consequences of their actions might help in getting them to stop for awhile but it would be a better tactic to encourage parents to connect with their kids more, build a better family life and provide them with real education about how to monitor their teen's cell phones so it is not abused. I've noticed the same officer giving plenty of media interviews about what's been happening. He's already spoken to Boston TV Channels 4, 5 and 7. The officer is the police department's lead guy for the investigation since he's the expert on crimes against children on the internet. I've got to say that I don't agree with the officer when he says that the child pornography charges the boys face apply to their case. Then later he says the opposite and talks to reporters telling them "the sex offender laws weren't made for the teen sexting scenario." Why throw the heavy hammer of the law at the kids when it doesn't apply? Here's my rant again! The laws should be changed to apply to minors; after all, a minor offender is not a major offender. There's no way you can put an adult who intentionally creates, views and sends photos of naked children to satisfy his lust and perverted pleasure in the same category as a teenager who responds to peer pressure, adolescent hormones, and possibly to an underdeveloped prefrontal cortex.

I learned about an underdeveloped pre-frontal cortex from taking a look at the research of Dr. Jay Giedd at the National Institute of Mental Health. Dr. Giedd says: "It's not that the teens are stupid or incapable of (things). It's sort of unfair to expect them to have adult levels of organizational

"...*under surveillance*"

Skills or decision making before their brain (prefrontal cortex) is finished being built." Dr. Giedd sounds like he has a clue about what he's saying, but I don't know that I believe him. I've learned that making excuses for problem behavior isn't usually a good idea.

The other day I saw the same officer in stake out mode. He was sitting in his car spying on students while parked across the street from the school. I can't imagine he's seeing a whole lot of criminal activity. What does he expect the teens to be doing? Sexting? He would need a super telephoto lens to catch any of that going on.

The more I find out about the whole sexting crisis, from the way it's being handled by the school authorities and police department I have one word for it: overkill. Ants are being blasted by nuclear weapons. Better go and make my bologna sandwich, I'm really getting hungry seeing all the bologna that's been flying around here.

Blog Entry 7 of 10 – After It Happened
Missed Court Appearance
Friday | February 13, 2009 | 4:45 p.m.

Yesterday was an important court hearing for the boys. I missed it. A friend of mine phoned and said "Hey Brian, I heard your name called out in court. Why weren't you there?" I told him. I didn't know I was supposed to be there. I didn't get anything in the mail notifying me of the hearing date or time.

I worried and thought that I could be held to a contempt of court charge and a warrant for my arrest would be issued. My wire was bent even more after a "Cape Cod Times" reporter called and asked me why I didn't show up in juvenile court this morning. I said. "No summons arrived in the mail."

What happened? I called the court clerk about an hour ago and found out that a notice was sent to my home address not my P.O. Box. I have a box at the post office because mail is not delivered to my home. I don't understand why this mistake wasn't caught. I was sure the lead officer on the investigation had my correct post office address.

I remember when I spoke with him a couple of days ago. I wanted to talk to him about something he said at the police station to my wife during the parent meeting. When he came on the phone, I asked my question and he said with a curt tone. "Mr. Hunt, your wife doesn't listen too well! I gave my answer to her at the meeting the other night

the library where Kim works

The officer's response surprised me. If I wasn't so blurry brained I would have said. "Back off bud, my wife knows how to listen well. The wax needs to come out of your ears, because you don't hear how rude you are. Kim has been the town librarian for almost thirty years and she knows what it means to listen. If you were in what you said, she would have heard you. Communication is a two way street." I wish I had said what I was thinking but I didn't. I did my best to be polite to "Mr. Rude," but my "clam" was pretty boiled!

Rude is inexcusable. Why people in power and authority go there is beyond me. I hung up the phone and felt angry over what happened. Maybe he's upset because he's not appreciating the fact that I'm speaking my mind to the media, and because of it, he's got to put a whole lot more time on the case.

I remember a time a couple of years ago when this same officer was directing traffic in front of my house. It was a hot day and I wanted to give him a bottle of cold water as he looked pretty warm standing outside on the pavement in his uniform waving his arms. I walked out to where he was and gave him a bottle of water. He thanked me.

What I've noticed about officers like him, is that they try their best to do their job, pleasing the community they serve and keeping the law breakers in check

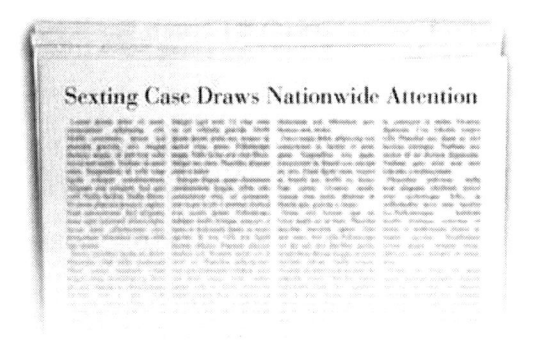

Sexting Case Draws Nationwide Attention

and I appreciate their efforts. But I must say that they really need to save the snappy bulldog's bite for the real criminals and not give it to a parent whose breaking sweat to do his best to protect his kid and others from a wrong application of the law. Politeness and respect go a long way. I'm no philosopher, but I do have my observations and opinions, and for what it's worth here's one. There are those who want everyone around to understand they're "in charge" and wear "the uniform," in whatever job they're in, be it a corporate executive or a lead janitor. Folks like me better not come up against 'em and bother 'em with stupid questions. Someone told me something about power not long ago and I really liked what they said. "Nothing delights people of weak character more than having false power over others. They get a kick out of causing people problems, being difficult and causing grief, the more they give of it the better they feel. They also like trying to get negative reactions from others. The best way to handle them is to not set them straight, but to walk away. Besides, they wouldn't want to hear it anyway." Ben came home from school today telling me that most of the guy students had the girl's picture on their cell phone but they deleted it because they were afraid that they might get caught and be down at the police department too. Ben said that a lot of the kids were bugged that the officer had been parking his patrol car across from the school in stake out mode trying to catch students doing wrong. Why not catch them doing good and giving positive encouragement? How about giving teens a "Way to Go" award when they do something worth congratulating them for?

    Ben was proud to tell me that his friends thought he was taking the whole thing pretty well. He told me, "You're a cool dad to be fighting for me and the other students too!" Why wouldn't I give it my best shot? Ben's my son. I'd give up my life for him if I had to. I gave him one of my best father hugs, the kind where you wrap your arms tight around and hold for a few seconds. I don't think Ben minded one bit!

A producer over at "Inside Edition" phoned a couple of hours ago. I told her "Ben and I would like to go on the show." The truth is the thought of going on national television makes me nervous but I'll get over it, because I want my story and frustration heard all over the country. Ben said he would be happy to do what he could and that a trip to New York wouldn't be so bad. One of the other dads and his son will go on the show too. He told me one of the reasons he'd go would be for the free trip to New York City which would include a complimentary stay at a five-star hotel, free limousine service and $675 cash for each of us to use as spending money. We're going to leave next Friday morning on the twentieth and tape the show later that day. We'll have our weekend to play and fly back on Monday. The show will air on Tuesday, and then the next big event will be my hearing on Thursday. This time the court sent someone over to my house and the summons was hand delivered.

Blog Entry 8 of 10 – After It Happened
Agreeable and Disagreeable Opinions
Saturday | February 14, 2009 | 9:30 a.m.

Bloggers are blogging about sexting with the passion that a graffiti artist would have spraying a large cement wall with a message. Opinions are like grains of sands on the seashore, too many to count. I'm not ready to go on-line with my opinions. You could say that penciling my thoughts as they come is better for me than typing on a keyboard. Never did well in typing class. Average ten words a minute. Editorials and letters to the editor are popping up in local papers. Some make me mad. I'm amazed that those who don't have kids are the quickest to criticize, and tell me how I should have done it differently. I'm glad I read a good letter or two from the several sent to the editors. Here's what I remember from one of the better letters.

If there was a kindergartner who had a cell phone they would likely take a picture of themselves going to the bathroom or pulling a booger out of their nose.

So, a 13-year-old girl pulls up a corner of her shirt; a 14-yearold boy takes a picture with his cell phone and then he sends it on to a friend; who then does the same. Now six boys have the girl's picture on their screens, maybe hundreds of other school age boys have it on their cell phones too.

Has a horrible, terrible, evil crime been committed? Such as possessing and distributing child pornography? No! Of course not! Not by any stretch of the imagination! Such a crime would be committed only if someone stupidly applied a very strict legalistic application of the law. If they did this then they would be wrongly justifying sending teen sexters to jail! Here's a letter I also remember but this one I don't like. "Old Pop just doesn't get it? Hunt said he had to buy his son another cell phone because the authorities will hold on to it for months and he likes to know where his son is at all times. Old Pop's kid has shown that he can't manage a cell phone and after all this mess, shouldn't Junior be home doing homework and chores!"

The letter writer doesn't want to think about the idea of a father who cares about his son's safety and wants to know how to reach him. A cell phone is very important in an emergency. Oh, by the way, Mr. Letter Writer, my son already does homework and chores and has since he started school at five.

Blog Entry 9 of 10 – After It Happened Reflections
Sunday | February 15, 2009 | 9:15 a.m.

Today is Sunday. Not a church guy, but feeling reflective. Going to St. Patrick's and lighting a candle may not be a bad idea. But doing that, doesn't feel as real as lying back on my pillow at night and just spilling out my frustrations to God, after all if He's the one that put me together then He shouldn't have a problem in understanding what I'm going through.

The hardest thing to deal with in all of this is my worry about Ben being charged, being found guilty and having his life ruined. Justice doesn't always play out with the law or the courts. When I look at my wife, I see fear in her eyes. He's our only son and she can't stand the thought of her baby having a felony record at thirteen. She's scared, I'm scared, Ben's scared, and we're all scared. I think I'm going to have a word with God. "Help!"

Calling out to God gives me a feeling that everything will work out how it's supposed to. I'm asking Him for wisdom, the understanding to do what I need to in the right way at the right time. It's not knowing now how it will all exactly work out that bugs me. I have to tell myself, over and over. "I'm not in control, I'm not in control." In one of my more brilliant moments, I think "why would I be in control, when I don't see very much." After all, I only look at life through a tiny peephole. My vision is beyond small compared to the big eye of God. I should just trust Him and shut up. This I know, my fears and worries can wrap me pretty tight, like I'm rolled up into a sleeping bag, without air to breathe. Just heard someone remind me of the obvious, no one controls their heart beating, nor does anyone control their lungs breathing. It's kind of humbling when you think about it. The most important organs in my body that keep me going, I don't have a bit of control over them doing their thing, at the perfect time, like beating my heart, and giving

"...reliable and steady"

cruise down our coast, and enjoy the mystery and beauty of the ocean and we'll stop and look at the old tug boats in the harbor. They're one of my favorites, certainly not pretty to look at but they remind me of how important reliable and steady are. Once we're home, I'll put on one of my clam boils, invite a few friends over and do my best to shoot some hoops with Ben, I'm still in pain from the surgery. Kim will light the candle for us later tonight. I don't want to get so caught up with the whole thing that I lose sight of what's important: God, my family, doing good and helping others whenever I can.

Blog Entry 10 of 10 – After It Happened
Down and Up
Monday | February 16, 2009 | 9:03 a.m.

I'm feeling down today. I'm reading too many negative opinion letters, like this one from the morning paper. This mom rants and steps up on her self-righteous box. She talks about how parents should teach kids to be responsible and how they should set the rules on when their teen can use a cell phone. She goes on to say that if a parent teaches their kid cell phone responsibility and he abuses it, then shame on him but if you've haven't taught it and he gets in trouble shame on you.

The mom doesn't understand that if a parent hasn't heard about sexting then they wouldn't think about telling their son or daughter about the dangers of doing it. Education is important. Got to thinking about this the other day and it's worth writing down. If it's technologically possible it would help kids and teens if there was a warning message on their cell phones that pops up on the screen before they use the camera. The message might say:

Annie Winston

> Dealing With Teens, Sex and Cell Phones

"If you are younger than eighteen years old and sending 'racy' photos as a text or email attachment to someone your age or younger then you are breaking the law and committing a felony. If convicted you'll be found guilty of trafficking child pornography and possibly be going to jail, having a sex offender label after your name for forty years and paying a lot of money for a fine. Think before I just finished reading another letter but this one isn't full of self righteous why did you not do it as good as me attitude. It's from a couple of school media educators from Cape Cod. I like it a lot! I'll put what they said in my words. The red light district flashes its red re you click and send. Sexting is no joke!"

light through cables and the airwaves, like an uninvited guest it rudely arrives into the privacy of our homes. Our families are at risk! Messages and images come in through the living room TV that should only be on the adult X-rated channel. If our media is the main source of storytelling and a major tool that communicates what people value and want in society, then why should children be punished as sexting felons for only copying what society says is "good and acceptable" and openly communicated through television, commercials and movies and halftime shows at football games. (I remember Janet Jackson's accidental exposure of her breast on national television during a Family Time Super Bowl Sunday a few years back). Only blaming parents and teens in sexting incidents is not fair because once a child leaves home and ventures out, they come under all sorts of unwanted influences.

Once outside the home kids have to fend off hell's assault on their purity because there are no protections in place to protect them. How about creating new laws and policies about what should be or shouldn't be on the screen and airwaves? It's not censorship but simply PROTECTING our children. Come on, as a society, we've become Rip Van Winkle, and fallen asleep! Our precious kids are suffering and they are paying the price of not knowing right from wrong, and they're living in a moral slime pit where anything goes. Now, I'm inspired! My down is gone and up I am! Press on and fight, I will!

Blog Entry 11 of 20 – Media Frenzy
A Father's Heart
Tuesday | February 17, 2009 | 10:17 a.m.

In a few days we'll be boarding a plane and heading to New York to go on "Inside Edition." I'm not feeling as nervous as I was, but nervous enough. There are times when I wonder if this whole thing is just a bad dream. I have learned to stop asking the question: why did this sexting scenario have to happen to my family?

I know one day truth will win out if I continue moving forward trying my best and struggling to do what's right. I believe that I won't be led down the river and over the falls. I want what's fair for my son and he getting slammed with adult sex offender charges is definitely not fair. I'll fight until it's dropped or I drop because my father's heart won't stop beating.

Blog Entry 12 of 20 – Media Frenzy
Mouthing Off
Wednesday | February 18, 2009 | 8:45 a.m.

There's a snow storm raging outside and it mirrors how I'm feeling about this whole sexting issue, whipped, buffeted and in the blinding white. I stare out the window and see swirls of snow spiraling in the air looking like little spinning corkscrews. I continue to lay awake at night worrying and wondering why the whole thing happened the way it did. My rant won't leave me! Teens sending nude pictures over electronic devices should not get the same punishment as adult child pornographers. I need to stop and catch myself from going into the world of "shoulds." It's too easy to go there and it doesn't make anything better. This I know, life certainly has its storms and it seems like all of Cape Cod is following mine. Can't go anywhere without someone coming up and saying "Hey Brian, saw you on the news last night. Keep on fighting! I'm proud of you!" These folks are on my side but there are still those letter writers who are not and it's those types who get me going because they don't care about a father whose trying his best to keep the law from cutting "a pound of flesh" from his son. Here's my version of a not so friendly letter to the editor. The writer quotes the Bible. "Do unto others as you would have them do unto you." Then he lays into me for not teaching Ben about compassion and empathy for the girl who sent the photo. Does the letter writer have a secret camera recording what went down with Ben and my discipline of him? Of course not! The letter writer has no clue into my son's heart or mine. Ben has plenty of compassion and empathy for the girl as I do. He feels sorry for the girl because she didn't think too much of herself.

My compassion for the girl starts with my action to protect her by not talking about what she did and making her the problem. My mom doesn't agree she says "the one who started a problem is the one responsible for the problem." My son clearly knows what he did in passing along the girl's photo was wrong and now he knows how very naïve he was in doing it. He did not understand that he was breaking child pornography laws the moment he clicked the send button on his cell phone and sent the racy photo to a friend. I want to say that I also feel bad for the girl. She victimized herself and probably didn't know it. Teens who flash themselves are teens who don't respect themselves. They've believed a lie about how they are going to be more cool and desirable if they show it. Cool and desirable they might be, but to the wrong types those that would end up hurting, scandalizing and dumping them.

Now I am going to do some serious mouthing off. The father gets on my case even more by saying something like this:

"He has spun the sexting problem in a way to make his son

look like the victim." Let me say this: I didn't spin anything! Ben is a victim of unfair laws that weren't written for teens who act out of stupid impulse and send a photo they should have never sent. Both Ben and the girl are victims, and so are lots of other teens, who are lied to every day by the media. The lying message sent, each day, in thousands of different ways is "take it off, be sexually free, and forget about keeping it on and protecting one's purity."

'Sexting' Controversy in Falmouth

Teens are blasted twenty-four seven by the media with "be sexually free" messages in subtle and blaring ways, just like the school media educators talked about in their opinion letter. Such messages flow into teen's minds and hearts like a steady stream of water from a faucet that won't turn off. The flow comes from the music they listen to, the reality shows, movies, television programs and commercials they watch, and from the books, magazines and newspapers, they read. Teens live in a world that has sexualized everything even something as normal as eating a hamburger. The lie of "be sexually free and you won't get hurt" says, "teens have all the fun you want! Take it off! Encourage others to take it off! Don't worry there's no price to pay of shame, embarrassment or humiliation or feeling like you want to die, because everybody is taking it off and making jokes about it too. Join the party don't keep it on and be pure, that's no fun! Purity is for prudes!"

Teens are victimized by the lie and when they live it out in their personal lives it often comes at a very high price. Just ask the mom of Hope Witsell, a thirteen-year-old girl in Florida, who "took it off" and sent a bare breasted photo of herself to her boyfriend. "Somehow," a girl student got a hold of Hope's picture and forwarded it on to students at six different schools. Before long Hope's racy photo became viral and the bullying intensified. She was called "slut," "whore"' and "skank" and a cruel cyber bullying campaign was launched with Hope as the target. Hope didn't open up about her emotional pain to family or close friends instead she hung herself. The true message the media needs to send to teens is: Don't take it off, keep it on! Purity is freedom and life and the opposite is not. Better go pick up Ben. It's likely that the officer will be staked out in his car. I wonder if he's got a pair of binoculars up to his eyes this time. It burns my toast that he's still parking out in front of the school like he's a "Mr. Going to Get You." It would be better if he took up bird watching instead of student watching or how about if he got out of his patrol car, walked around the campus being a "Mr. Howdy Doody," one of those super nice and super friendly officers who reach out and really care about kids in the community. I've learned that kindness and friendly go a long way, especially with teens.

Blog Entry 13 of 20 – Media Frenzy
Coffee Musings
Thursday | February 19, 2009 | 10:17 a.m.

Here with a cup of coffee thinking about the trip to New York. Can't believe we're leaving tomorrow. Can't believe how incredibly patient my wife has been through this ordeal. Kim's the love of my life. We met in high school and we've been sweethearts ever since. She trusts my judgment and tells me. "Brian you just handle the whole thing because I know your heart is in the right place about it all." She's right. A friend once told me, "a wise guy listens to his wife." I listen to my wife, so I'm a wise guy. I've learned that with if my heart is in the right place, my feet will follow and everything usually works out pretty well. I read another letter to the editor in the paper this morning. The sender was complaining about the sensational news reporting. I can't

Me with Kim — "the love of my life"

remember everything he said but this is what stuck with me: The sensational reporting of what the boys had done and the way the police handled everything was over the top, to rush to judgment and threaten the boys with charges of pornography, jail time and a lifetime label of "sex offender" is crazy.

I want to shake this letter writer's hand, I agree with him. But the truth is that bringing down the gavel on the students didn't start with the newspaper reporters (they were only reporting it). It likely began with the school authorities and police whose "heart may have not been in the right place." It's easy to rush to judgment especially when pride and fear fuel it.

Blog Entry 14 of 20 – Media Frenzy
To New York City
Friday | February 20, 2009 | 10:17 a.m.

I'm at the airport about to board the plane. "Inside Edition" and New York City here we come. I kissed Kim good-bye right before we left and told her that I loved her. You can never tell your loved ones that you love them enough. You might not see them again. The plane could crash. A terrorist might blow it up with a shoe bomb, or it could go down because an airplane mechanic leaves a wrench in the wrong place, or it

away we go to New York City

could hit a flock of birds. I remember the US Airways pilot, Captain Chesley "Sully" Sullenberger, who hit a flock of birds and was able to land his severely disabled jet with 155 passengers on board on the Hudson River after a couple of geese got sucked up in both of the plane's engines. Kind of interesting and strange but the day of the US Airways incident was January 15, 2009 and it was the same day that the sexting investigation began at Lawrence Junior High School.

I have to say, Captain Sullenberger is my hero. His character is strong, straightforward, and super cool under pressure. This is something I greatly respect and admire. Just heard our seat section called and I also hear Ben and John Jr. shouting "Don't leave without us!" I warned them about taking a quick trip to Krispy Kremes. Going anywhere last minute never turns out to be a "quick trip." There's a lesson here in the dash for a donut. Don't run for the donut dashes or anything else that might distract you from a goal, be it big or small, like boarding a plane on time. What I'm saying is really not bad advice for life. Don't get too distracted by the little stuff that comes your way it just might cost you the bigger stuff.

**Blog Entry 15 of 20 – Media Frenzy
A Star I Am Not Saturday
February 21, 2009 | 6:30 a.m.**

I'm catching the early bird worm this morning. We're staying at a plush pad. It's a five-star Times Square hotel, where chandeliers dangle from the ceilings and lots of overstuffed furniture is in the lobby making the place appear like a thirteenth century castle in medieval Europe. I'm really enjoying how the nicely dressed staff are always asking if they can help you. Feeling like I'm some big shot from a large company.

A star I am not. I'm just a father who loves his son and

wants to protect him from laws that weren't written for his "crime." Little does the hotel staff know that I'm here because the tab was picked up by a news show who was interested in a story about a father's sexting teen. I should go back to sleep but I can't. I'm too excited. I want to get on with the day, touring the city, having fun and buying my wife a souvenir. I don't think she'd like a coffee mug with a picture of the Empire State Building. Ben is still sleeping in the bed next to mine. I can't help but get a bit choked up. When I remember the day Kim brought him home from the hospital, small and tightly bundled, and how proud I was to be his father, and still am. Nothing will ever take that away. My love for Ben won't change no matter what he says or does. I want the best for him. That's why I won't give up the fight until I know for sure he's not going to be wearing the orange prison jumpsuit. What a great day we had yesterday. Starting with our plane landing safely at the airport I was happy that the engine didn't suck up a flock of birds and the pilot wasn't forced to land on the Hudson River. A nice guy with a five-star limo service picked us up at the airport and we were taken to "Inside Edition's" studio. The staff was sympathetic to what we were going through, especially one of the producers, who gave us her best efforts in helping us feel comfortable and relaxed. She escorted us into one of the Senior Producer's offices who had a huge window overlooking the Hudson River. One of the staff told us that everyone at "Inside Edition" had a front row seat watching Captain Chesley "Sully" Sullenberger land the plane. The staff guy told us how amazing the sight was to watch and how helpless everyone felt standing there seeing it happen knowing that the likelihood of anyone coming out alive would be very small. I'm glad the drama had a terrific ending. I sure hope mine does too.

A short time later, one of the producers walked into the office and said that it was time for our interview. We went into a small sound recording studio that had been completely blacked out with thick dark curtains on the windows. We took a seat and were wired up. Headphones and microphones were put on each of us. Sitting in the studio wired up (this time without coffee), I felt overwhelmed when I thought how I was going to be seen by millions of people all over the country. The truth is I didn't really care how overwhelmed I felt or whether I stuttered or slobbered. I want my story out and my frustration aired. As I've said, public outrage over what could go down on my son and the other boys might be stirred up and further pressure put on the police department and court to lighten up on the kids.

The kindly staff at "Inside Edition" continued to reassure me that they were "behind me one hundred percent" as they had children the same age as Ben and John and if they were in my shoes they would have done the same thing.

After we finished in the studio, we were told that camera crews would be filming the boys at a basketball court and bowling alley not far from where we were. They wanted to show Ben and John doing normal fun activities that teenagers do. I'm all for normal fun activity as long it doesn't get them into trouble. Heard something awhile ago that is very true. "Man is born into trouble just as sparks fly upward." That's a quote from somewhere in the Bible. Believe me, I'm no Bible thumper but I thought this was worth writing down, because teen sexting is nothing but heartache and trouble.

Annie Winston

Blog Entry 16 of 20 – Media Frenzy
Walking, Walking, Walking
Sunday | February 22, 2009 | 9:30 p.m.

The day whizzed by. Between yesterday and today, we put some serious wear on our sneakers. Walking, walking, walking were we, probably logged in over twenty miles between the two days. New York has so much to see and we didn't want to miss one sight. There's Times Square, Central Park, Grand Central Station, Wall Street, the Rockefeller Center and a whole lot more. We toured the city taking in everything around us from the horse drawn carriages, to the Rickshaw Pedi cabs. Ben wanted to ride one so we did. I thought the driver's bicycle tires might pop with the amount of weight he was pulling with us in the cab.

For lunch we went down to the lower east side of Manhattan to Chinatown and sat down at a small Chinese restaurant with tables outside. I was surprised to learn that there were two hundred Chinese restaurants in Chinatown. I studied the menu and read over it carefully and became a bit queasy thinking about how Hairy Gourg Egg Soup might taste. I passed on that one and chose Szechuan Beef Stir-fry. I could have chosen Fish Head Curry, Crab in Fermented Soya Paste, Shark Fin Soup, or Braised Abalone with Sea Cucumber. Despite the exotic menu items, I have to say that

I'm impressed with the variety, color, flavor and aroma of Chinese food. I never forgot what someone told me one time something that Confucius, the Chinese philosopher said: "The path to your friend's heart and soul begins from your cooking." I love to cook, and my friends know it as I'm always serving up my special Cape Cod dishes like scallops wrapped with bacon and broiled to perfection. After Chinatown, we took a bus and hopped a boat to Ellis Island and saw the Statue of Liberty. I learned that the length of the sandal on Lady Liberty's foot is 25 feet. I'd hate to go shoe shopping with her! Speaking of shopping, I bought Kim two souvenirs. A green scarf and a snow globe of the Statue of Liberty. I hope she likes them. Time to turn the light out and go to sleep. Ben is already snoring. This I know, I'm glad to be leaving tomorrow. New York is exciting but home is better.

Blog Entry 17 of 20 – Media Frenzy
Not Something to Brag About
Tuesday | February 24, 2009 | 1:00 p.m.

Kim liked her green scarf and snow globe. She's got the globe on the fireplace mantle next to the clock. She's counting the hours and I am too, when "Inside Edition" will air our show. I plan to pop popcorn and make it a real debut, a grand one! Everyone in town knows about it and most see me as the parent's megaphone on the sexting issue because no one else really wanted the job of standing up to defend the kids and having to admit publicly that they are a parent of a sexting teen. It's certainly not something to brag about to the relatives at Thanksgiving dinner.

I'm willing to battle to protect my son and the other kids because their future hangs in the balance. I don't need to be a space engineer to figure out that if a sex offender conviction sticks, it will be impossible for Ben and the other five students to get into college, serve in the military, get a decent job, and find housing or anything else worthwhile that will help them move forward in life. I recently read that teens who send a racy photo of themselves or receive one from a teen girlfriend or boyfriend and are caught by the law will have a tougher penalty to deal with than other hands-on type sexual offenses. The teen will be forced to register as a sex offender for ten years or more. The Federal Adam Walsh Child Protection Act of 2007 requires that sex offenders as young as fourteen register. This is why I am willing to put my face in front of the camera and do everything I can to fight the charges against my son and the others. Lots more calls are coming in from the media for interviews.

"Nightline" is interested in doing a story. "Oprah's" producers have called. A writer from "People" Magazine wants to talk and "Dr. Phil's" producers have left messages for me to call them. I'll go on every one, five hundred times, if it will save my son's life.

Blog Entry 18 of 20 – Media Frenzy
No Worries Wednesday
February 25, 2009 | 10:00 a.m.

Our segment on "Inside Edition" aired last night. For me, it wasn't so grand. Big I was, but I'm happy to say that seeing myself didn't bother me this time. A work in progress I am, and there will be a day when the weight comes off. Kim thought I looked handsome (but she's my wife) and Ben said that we didn't look too bad. Each of us came across clear and strong with our message and it was a very good thing that not one of us slobbered or stumbled on our words. No worries will

I have about turning down offers from Hollywood to star in a new film. But Ben may get the offers and when he does, I'll be

his loudest cheerleader! I'm still being rammed on blogs for giving so many interviews to the media and some are still blabbing about how I'm seeking my own interest. That's nuts! A guy my size, doesn't care to be in front of a camera. Raising awareness about sexting and my son's case is what I'm up to and I won't apologize about using the media to get the word out. What father in their right mind wants their son registered as a sex offender?

Blog Entry 19 of 20 – Media Frenzy
Small Town Wednesday
February 25, 2009 | 4:30 p.m.

I went to town earlier today and people were coming up to me saying, "Hey, I saw you on "Inside Edition" last night! Good job!" It seems like everyone in Falmouth has been tracking the story wherever they can find it, be it TV, internet or newspaper. My childhood buddy, Chris, who lives close by, called me up the other day and said, "Hey Brian, saw you on TV when I was in Mexico." It's too true that life in a small town is where everybody's business is everybody's business.

Falmouth is a part of Barnstable County and lies on the Southwestern tip of Cape Cod. I've lived here my whole life and right now I'm sharing the 44.2 square miles of land with 32,000 people. I'm proud of where I live and I love to talk about it to anyone who'll listen. Falmouth is bordered by Bourne and Sandwich to the north, Mashpee to the east, Vineyard Sound to the south and Buzzards Bay to the west. Martha's Vineyard, an island, is about 3-1/3 miles north-northwest of us. Falmouth is the closest land to the island. We've got a lot of small ponds, creeks and inlets surrounded by tall pines and oaks. Our South Shore has quite a few ponds and rivers spaced close together and they travel into the town.

Our beaches are a bit on the rocky side. So ships and other vessels don't crash into the rocks, we've got the Nobska Lighthouse in

Nobska Lighthouse – Cape Cod

Woods Hole that was first built in 1829 on a small bluff facing out towards Vineyard Sound. It sits high up warning ships with its light and deep fog horn. In the first year it was built, it's been said that more than ten thousand ships went by safely. Falmouth has the only terminal for the Nantucket Steamship Authority that ferries to Martha's Vineyard. I've picked up many a passenger there for my limousine service. I always like chatting about the history of Falmouth to my riders if they're willing to listen. I usually tell them that Falmouth was founded by English colonists in 1660 and it was named by the explorer Bartholomew Gosnold for Falmouth, England, which was his home port. Sheep husbandry, farming, salt works, whaling, and shipping went on in the early days of the colony. By the early 1800s Falmouth averaged fifty sheep per square mile and by the late 1800s cranberries and strawberries were being raised for the market in Boston which is about eighty miles away. Falmouth's first train came to Woods Hole in 1872 and then people started building summer homes. Falmouth almost had some brief military action in the War of 1812. Falmouth Heights was surrounded by several British ships and the Massachusetts militia positioned themselves on the beaches to prevent them from landing but no shots were fired. For all the anticipation that something serious would happen, nothing ever did. That's a good thing though not very exciting. I also like to tell people that Katherine Lee Bates was born in Falmouth in the year 1859. She's the one who wrote the words to America the Beautiful. America is beautiful, but I must say that what's been going on with teen sexting and how the law is being used as big club to pound the teens, is not beautiful. Can't imagine what a great writer like Katherine Lee Bates might say about it all, especially when she wrote such inspiring words in her famous song

"God shed his grace on thee and crown thy good with brotherhood from sea to shining sea!"

Blog Entry 20 of 20 – Media Frenzy
Nervous Wednesday
February 25, 2009 | 9:30 p.m.

Tomorrow is the hearing. I'm nervous. I want the whole thing over with. I'm going to bed early and I'll be talking to God and ask that He shed some of that grace, Katherine Bates wrote about in her song America the Beautiful.

Blog Entry 21 of 30 – Everybody's Talking Can't Sleep
Thursday | February 26, 2009 | 4:00 a.m.

Can't sleep. Talked to God but before I did, I stared at the popcorn ceiling longer than I wanted to. I couldn't help myself. I finally fell asleep then woke up a few hours later and I've been awake now since midnight and I'm here sitting at the kitchen table penciling my thoughts.

Don't want coffee, so I'm having warm milk. I read somewhere that warm milk is what you should be drinking if you can't sleep. Melatonin is in it and it's a natural chemical that helps you fall asleep. It sure would be nice to be sleeping like a baby right now but I keep imagining myself in the courtroom standing before a stern judge who doesn't have the word "mercy" in his dictionary. I look over at the kitchen clock and I'm six hours from walking into court. The media will be swarming like I'm a King Bee (if there was such a thing). I want to come up with a creative idea of how to make my point about the charges against my son and the students. Maybe if I fall asleep and dream like a baby, I'll wake up and know what I want to say. I'm going back to bed now, feeling sleepy, maybe the melatonin kicked in.

Blog Entry 22 of 30 – Everybody's Talking Like a Baby
Thursday | February 26, 2009 | 7:00 a.m.

Alarm just went off, and I slept the last few hours like a baby. Drinking warm milk worked. I'm feeling refreshed and ready to brace for whatever impact comes my way today. Still don't know what I want to say to the reporters after the hearing. "Like a baby," "Like a baby," keeps coming to mind. I've got my answer! My VISA gift card has a picture of a naked baby on the front of it. Yet, no felony charges are being filed against the photographer who took the picture and I am certain that bankcard holders are not being held accountable for possessing a photograph of a naked baby in a flower tulip. My comparison doesn't line up one hundred percent but it lines up enough. The point is it would be crazy to charge the baby photographer and the holders of the card with a felony just like it's crazy to charge my son and the students with felonies. I'll say to the rolling cameras and flashing bulbs, "I cannot believe that a naked baby photo was allowed on this gift card."

Blog Entry 23 of 30 – Everybody's Talking
Hurry Up and Wait
Thursday | February 26, 2009 | 3:00 p.m.

Ben was running late this morning before we left our house for court. I hurried but had to wait for him to finish showering and getting dressed. I have a saying, "a teen in the shower is a teen in for an hour." We got out the door and only had 20 minutes to arrive at court on time. I pressed the pedal to the metal and blazed through a yellow and got to check-in with two minutes to spare before our 10 a.m. hearing. Ben and I were frisked by security, but the beeper went off for me. I forgot to take the coins out of my pocket. Once through we sat down outside the courtroom and waited. Again, I hurried to wait. Whenever I deal with court appointments, doctor appointments or DMV lines, I am filled to the brim with hurry up and wait moments. I'm learning to make the most of the time, not wanting to waste any of it, because life happens between the deadlines and appointments. Ben told me how much better it would be if this whole thing was over and we were waiting in line to get on a ride at "Universal Studios." I couldn't agree more!

I had a crazy thought in my "hurry up and wait moments"' that the bailiff might grab Ben by the scruff once his name was called and throw him into a paddy wagon and haul him off to jail. I'm glad that didn't happen. Once we got into the courtroom, we waited some more until the Magistrate announced Ben's name. He stood up and answered the Court's questions and then sat down. The exchange took about three minutes. I've come to a conclusion, though not a very insightful one. Much of life is hurry up and wait. We hurry to get there and then we wait. Now, I have to wait again on the Court and the US Postal Service. I really wanted a resolution in the case but I didn't get one. Instead I have to wait for a notification in the mail about what the next step will be. There is nothing I can do to hurry the answer. A lesson I've learned, life happens while you're waiting for an important something (whatever it is) and I would be wise to make the most of the moments while I'm waiting, because oftentimes what I'm waiting for isn't always as interesting as what happens while I'm waiting.

Outside the courthouse, the media crowd was buzzing around with trucks, camera crews and news reporters as they waited for the "Brian Hunt to say something meaningful and interesting moment." The truth is, I am not an important person, don't hold a big position at a company, didn't go to college, and I don't really have much to say except that I want my son free from felony charges and I want to warn parents and teens about the consequences of sexting.

I didn't want the media folks to wait any longer for me to say something, so I pulled out my VISA gift card and held it up to the camera. With Ben standing beside me, I said. "I cannot believe that a naked baby was allowed on this gift card." I waited for a response. It was quiet. I think they were surprised that I even said what I said.

You could say that my comment was as well liked as a can of flat room temperature Seven-Up. I moved on. "We came here today to put this all behind us, but we didn't get that. We have to wait until a letter arrives in the mail. I was told that it will take about a week to get. I really don't know about the outcome, but this I know, the whole thing has made everyone in my family, including me, very nervous."

I turned around and left with Ben. A few reporters asked some questions but I didn't feel like saying much. I wanted to go home. What more can I say? I'm just a father fighting for my son's freedom. One reporter didn't want me to leave; he followed Ben and me and asked. "How is Ben taking it all?" I looked over at Ben. He didn't want to talk. I said. "He's doing fine; we just want the whole thing to be done with. We're all hanging in there. We're hoping and praying everything will be all right. Now we just have to wait." Wait I will. At least this time I'm not hurrying to wait.

A Father's Sexting Teen

Father Seeks Resolution of 'Sexting' Case

Blog Entry 24 of 30 – Everybody's Talking
Proud Friday
February 27, 2009 | 9:00 a.m.

I've been reading the news articles about yesterday's hearing. Everybody's talking and has their opinion about what the outcome should be. Most agree throwing the book at the boys is not the answer. I was surprised to learn from the "Cape Cod Times" who reported yesterday that it was a couple of Lawrence Junior High students who came to the school counselor and said that a bare breasted photo was being passed around of a thirteen-year-old girl. Why my son and five other boys were "chosen" out of hundreds of students who had the bare breasted photo on their phone, I don't understand. But this I know, I'm proud of the students, whoever they were for coming forward to a school authority and finding the courage to say that an "inappropriate photo" was being sent around the school campus on cell phones. It's not easy to tell the truth especially when your peers might not think you're cool.

Blog Entry 25 of 30 – Everybody's Talking Vultures
Saturday | February 28, 2009 | 6:30 a.m.

It's early this morning. I wanted to do a little writing before I flip some flapjacks for Ben and Kim. I've got some thoughts I want to get out. I was thinking about them last night when I couldn't sleep. I'm still feeling jabbing pains on the side where I was cut open. They've been coming and going since I've been home from the hospital. Sure hope nothing serious is going on. Here's what I was thinking about last night. I wish more teens would stand up for what's right especially when it comes to reporting sexting and bullying. It's got to hurt and feel pretty awful if you've sent a naked picture of yourself to someone you thought could be trusted to keep it a "secret" and then before you know, it's being passed around to your peers like free water bottles. Then you feel even worse when you become the bullies and gossipers choicest morsel. You're now a student pin-up, an easy target for sly looks, crude cackling and cutting words. Can't imagine besides feeling like you want to escape and run to an igloo on Antarctica, what it would be like to walk down a school hallway enduring glaring eyes, and smoldering tongues.

I've noticed how mean spirited bullies hide out in the form of "hip" teens. "Hip" teen bullies should be called Vultures because they prey on the weak and vulnerable.

Vultures have sly mean looks, eyeing their prey, while waiting for the perfect time to pounce with a sharp beak and tearing claws. Teens beware: Vultures are real and they're deadly! Vulture attacks usually begin like this. They receive a nude or partially nude picture of a fellow student on their cell phones and start verbally attacking, either in cyber space or to the victim's face. The attacks show that the Vultures:

1) Get a sick kick out of being mean and hateful

2) Don't have the words, respect and honor in their dictionary.

3) Have little parent supervision or accountability at home.

4) Are jealous of their victim.
5) Are deeply insecure and they think that by putting another down they can feel better about themselves.
6) Don't believe that "kind is cool and a bully is a fool."

This I strongly believe, if a Vulture's victim doesn't get help, the wounds may be so deep that the teen despairs of living, and feels that the only way of escape is to take their life, like Jessica Logan did. Jessica was a sexting teen, living in Cincinnati, who sent a nude photo of herself to her boyfriend. After they broke up, he forwarded it on to a group of "gossip girls" who then started a cruel cyber bullying campaign against Jessica. Her photos eventually made it to seven area high school campuses. For weeks, she was bombarded with terrible texts, and evil emails. Jessica was mocked and called a whore, a porn queen, and slut. The poor girl put on a brave front, acting as if it wasn't getting to her, but inside she was ripping apart and in her despair and hopelessness she hung herself.

I read what Jessica's mother said to a Cincinnati Enquirer reporter and my heart went out to her. I feel her outrage and deep hurt. "My only baby that I will never be able to touch again. I will never have grandchildren. I will never be able to hand down my heirlooms. I'm just devastated by those parents that allow their children to do and say anything they want." Okay, I'm on a serious rant and I'm not stopping! When teens are loose with their lips and show a strong disregard for the feelings of another, it's usually because they have gotten away with it at home. Ben knows he can't get away with it in our home. Now he's aware just how hurtful sexting can be. Strong parent education about raising a teen to be kind, respectful and honorable towards everyone, especially his peer group, should be a part of every school's back to school night program.

Sexting and bullying are problems as difficult to solve as trying to get to the top of Mt. Everest with only a plastic spoon as a climbing tool. School administrators, teachers, parents, and teens unite to fight and don't give up! Disciplining bullies and making them accountable for their cruelty is the right thing to do. Other Vultures may think twice before they sweep in and attack their victim.

A teen who is constantly bullied for sexting or anything else, is a teen whose "cup" will shatter and break and they won't be able to hold anything anymore. The guys who write the psychology books would say that a person under great pressure and stress "will start to unravel emotionally and psychologically" and would be at risk for making the horrible choice of ending their life, just like Jessica Logan and Hope Witsell did when the Vultures sharp beaks and tearing claws became too much to bear. Bullied teens need critical life support of encouragement and help from trusted friends, school counselors, and family members. Let the good guys and gals, the "Supermen" and "Wonder Women" go after the Vultures on behalf of the hurting bullied teens. When the Vultures are identified and caught, school administrators must make them accountable to do what's right starting with giving a sincere apology to their victim as well as doing something nice for them!

Vultures should also do community service, attend a class with their parents on "How Not To Be A Vulture," and finally the captured Vultures should spend time picking up trash during lunch while wearing "I'm a recovering Vulture" t-shirt. Other Vultures who haven't been caught may stop if they see that there are serious consequences for trash talking another student.

I've ranted long enough. On second thought, not quite long enough. Some might wonder why is a guy who didn't go to college quoting a politician and statesman from the 1800s named Edmund Burke. I don't know but I'm glad that I came across his quote while surfing the net. I like what he says. "All that is necessary for the triumph of evil is that good men do nothing." I will say, as Brian Hunt, the father of a sexting teen, the husband of a good woman, and a limo driver living on the Cape, that sexting, and bullying prospers when good teens, good parents, good school counselors, good educators, good lawmakers, good local authorities, do nothing. I like what one school did when a girl student was being picked on in the cafeteria by a group of Vultures. They weren't guilty of doing nothing about bullying. When the parent of the girl called the school, their anti-bullying program kicked in. The school had a plan and it was activated. Each grade level had been organized into teams that were headed up by a teacher team leader. If bullying was going on towards a student in a particular grade, the team leader would go after the bullies and speak to them directly about their behavior and lay down the law. The bullies would be suspended if they picked on the girl again, or if they came up with new ideas of how to hurt her, or if they opened their mouth about the incident. Bravo to the administrators who acted quickly with a good plan when a bullying incident happened at school. The teacher (team leader) got tough and let the bullies know that their behavior was unacceptable and that there would be consequences. In the days that followed, the bullying ended and the girl student was no longer clawed and pecked at by the Vultures at lunch while she waited in the cafeteria line. The girl went on to be a thriving student doing well in school and making real friends and her mother was most happy. Her daughter had been protected by the leaders of the school and the Vultures were held accountable for their cruelty. This I will say loud and clear! Having a plan and uniting to fight against the Vultures really works!

Now here's my final mouth off to teens! "Good teens," if you know that bullying and sexting are going on at school or after school, on-line or off-line, and the "Vultures are closing in to consume their prey" then be a real friend to the one being "Vultured." Be courageous and report it to the school counselor or to a trusted and respected teacher, so that the problem can be solved and emergency help be given to the hurting teen. If you can stand up to the "hip teen bully Vultures" and you are willing to possibly endure their attacks without harm to yourself, then tell them to knock it off and stop using their beaks and claws to rip into another. Be cool, kind and strong. Be convinced that speaking truth and doing what's right always wins in the end even if for a time it seems like you're "losing." Dare to be a voice for the weak and defenseless. Dare to do what you can to stop bullying and sexting at your school. Dare to make your life count for good. Dare to live out Edmund Burke's words, "All that is necessary for the triumph of evil is that good men do nothing." Dare to challenge a Vulture from bullying but better yet, Dare NOT to be a Vulture.

Annie Winston

Blog Entry 26 of 30 – Everybody's Talking "I Wish I Hadn't"
Tuesday | March 3, 2009 | 10:30 a.m.

Been doing more online research about what teens are saying about sexting. I've pieced together a collection of random statements, taken from conversations of teens and statements I've read. What they all seem to be saying behind their words is "I wish I hadn't," "what a dumb choice I made," "how could I be so stupid."

"I didn't know I was breaking child pornography laws by sending out a picture of my privates to my boyfriend. When we broke up I was so humiliated and hurt when he sent my naked pictures to all his friends and then to make it worse, my photo was sent to all of their friends. I never thought my boyfriend in a million years would do that. He told me he loved me."
— **Teen Gal**

"My boyfriend wanted me to send pictures of myself in my bra and underwear. I did and now I wish I hadn't because he sent the pictures to his friends. I was so mad. I felt betrayed and now when other guys look at me at school I wonder if they have seen me.
" — **Teen Gal**

"Sexting is scary because once your naked photo is sent you can't take it back. It's always going to be there somewhere, even on the internet where some pervert on a porno sight could be looking at you."
— **Teen Gal**

"The biggest problem about sexting is you never know how bad the consequences could be. You might think you can get away with it, but that is stupid think because just when you least expect it, your photos could show up and get into the hands of the authorities and you could be at a police station facing child pornography charges and be labeled as a sex offender."
— **Teen Guy**

"Girls are hurting themselves whether they believe it or not when they send a naked picture of themselves to a boyfriend. They are saying "I'm a sex object, a blow up doll, not a person who deserves to be treated with respect and dignity
." — **Teen Guy**

"Girls think their boyfriends will like them more and if they don't send their 'photo' then they will stop liking them and they don't want to be left out, pushed aside for another girl."
— **Teen Gal**

"Guys want to be seen as sexually strong and collecting naked girls photos like 'trophies' is one way they try impress their friends. The truth is guys who do this are not being sexually strong; they're being sexually aggressive and treating girls like an object, like a Barbie doll or something. A wise guy respects a girl's dignity and purity. A foolish and selfish guy doesn't."

— **Teen Guy**

"I first thought sending the naked photo of a girl student at my school was kinda funny and just a joke and that nobody would really know. Now everybody knows. I have to go down to the police station and I'm facing felony charges for distributing child pornography because the girl was underage. What I'm going through is no joke!"

— **Teen Guy**

"Sexting is not romantic, cool or funny, think long and hard about the possible consequences, your life could be ruined."

— **Teen Gal**

"Girls, if someone is trying to pressure you to send a photo he's not worth being with you at all. Use your voice and say NO! Be confident in yourself. If a guy really likes you, he will like you for you! Not because you take your clothes off for him. Guys don't abuse a girl by asking her to take her clothes off and take a picture of herself. You're being sleazy, mean spirited and pathetic!"

— **Teen Guy**

"When a relationship breaks down, and a guy is hurt and angry about it, there's a good chance that he'll send out a naked photo that you sent him, as a type of revenge, to get back at you."

— **Teen Guy**

"I felt betrayed by someone who said they loved me. Once I knew he had sent out my pictures and they were being seen by everyone at school, I was sick to my stomach. Now I would be thought of as a "ho," and I would get a really bad reputation. I would be tempted to lie and say "no it wasn't me" and I know that's not right!"

— **Teen Gal**

"I just wanted him to like me more. I thought he would go and find some other girl if I didn't send him 'special' pictures of myself. He kept telling me how much he loved me and how hot he thought I was. I know now that he didn't love me, he "lusted" me. I've since learned that love is patient and kind. What he asked me to do wasn't patient, (I had only known him about a month) and it wasn't kind, (asking me to take off my clothes and take a picture of myself is not nice) and then pressuring me to do something that I really didn't want to do. My school counselor said sexual abuse is when someone forces either through persuasive words or direct physical contact to sexually act in a way you are not comfortable with and you know is not right."

— **Teen Gal**

"Sexting is a sure way of 'dissin' yourself, and totally 'dissin' your God-given dignity."

— **Teen Guy**

Annie Winston

Blog Entry 27 of 30 – Everybody's Talking Anticipation Makes Me Crazy
Wednesday | March 4, 2009 | 8:30 a.m.

I can't wait to check the mail. The letter from the court should be arriving today and I hope it comes today. Because I'm anxious, I going into extra cleaning mode, not only is the kitchen spotless, which makes my wife very happy, but my limo cars have all been polished and buffed to perfection. My anticipation of what the court's verdict is, makes me "crazy." I'm also on an extra dose of pain pills because the pain in my side still won't go away. I've called my doc and he says for me to come in and he'll check it out.

our spotless kitchen

Blog Entry 28 of 30 – Everybody's Talking
A Round of Applause
Friday | March 6, 2009 | 4:30 p.m.

I picked up the mail this afternoon and there it was, the letter from the Trial Court of the Commonwealth Juvenile Court Department Barnstable County/Town of Plymouth Division. I called Ben and told him the letter from the Court had arrived. I asked him if I could open it and he said "yes." Ben stayed on the phone and I read it aloud:

March 3, 2009

Dear Mr. Benjamin Hunt,

After giving careful consideration to the facts and the law with regards to the charges of possession of child pornography and dissemination for the same filed against you by the Falmouth Police Department, I have decided to continue both criminal charges until August 27, 2009. I will hold both complaints in my office and will dismiss the charges if there are not further criminal charges of a similar nature filed against you. Please be advised however that if the police or a private complainant has reason to file new

charges against you of a similar nature for incidents occurring after February 26, 2009, I will bring the old complaints forward. If I find at the new hearing that probable cause exists to issue the new charges, then I will issue the old charges (possession and dissemination) as well. If there are no additional criminal charges filed against you, then on August 27, 2009, I will administratively dismiss the present charges without you being present in court.

Very truly yours,

Charles P. Andrade Clerk Magistrate

...where I got a round of applause

Ben gave a shout. But then he quietly asked. "Dad, does this mean I'm totally off the hook?" I told him. "Only if you don't get involved with sexting before August 27 but that is not going to happen, right son?" Ben said. "I'm so over that stupid sexting stuff, messing up my life is not worth a cheap thrill or a laugh. It's no joke!" I told him that I was proud of him that he now understood the pain and stupidity of sexting! "Tonight Ben, we're celebrating! I'm buying steak for dinner and ice cream for dessert." I hung up and went over to the White Hen Pantry. I walked in the door eager to tell the employees the "good news." They've been following Ben's story. I said, "We're almost there! As long as Ben doesn't have any new sexting or other felony charges filed against him by August 27, then the charges are forever and completely dropped."

What happened next surprised me. The employees and a few of the customers standing around started clapping. They gave me a round of applause and the cashier came right up, looked me in the eye and said. "Brian we're proud of you, you hung in there and fought t! You didn't give up!" You didn't

...my praying Mom!

"Thanks guys! I couldn't have done it without knowing I had most of Falmouth behind me, you're great! Can't say that the media didn't help out a bit too! I haven't told you yet, but "People" Magazine called the other day and they're gonna do a story for their March 30 issue. "Dr. Phil's" producers are interested; I may be going on his show. But it's you all, your encouraging words that really put the "gas" in my car and helped me keep on fighting! I knew I had real supporters around town, except for a "few" letter writers!"

I left the store feeling the best I've felt in a long time. I could hardly believe the good news. I walked over to the library, where Kim was working and showed her the letter. She was thrilled and said. "Let's get something special for dinner tonight." I said, "I already have. I just bought steak and ice-cream at the White Hen Pantry. When I walked in the door, and told the employees the good news, I got a round of applause." "Well then, you get my round of applause too!" Kim started to clap.

At that moment, I don't think my rocket ship could have blasted any higher. The relief of knowing that Ben wouldn't be charged was wonderful enough, but to get the extra applause on top of it was like sprinkles on a plain donut. I got into my car and started to drive home.

I passed St. Patrick's church and remembered what mom said.

"God always answers prayer, just not in the way you think."

Now that's the truth! It was just a few weeks ago when I was staring at my popcorn ceiling when fear and worry gripped me like a tight tourniquet, and I said to God with no doubt, "HELP!"

Before I could even think about what I was doing, I made a quick left turn and turned my car around to make my way back to St. Patrick's church. I pulled into the church's parking lot, found an open space, turned off my car and did something I hadn't planned on doing. I clapped for God and gave Him a round of applause.

Blog Entry 29 of 30 – Everybody's Talking
What If Monday
|March 9, 2009 | 3:30 p.m.

Ben came home from school today with a question. "What if, somebody at school sends me a 'sext' message because they want to be mean to try to get me in trouble, knowing that I'm going to be completely off the hook after August 27?" I told him. "Son, if there's one thing you should know by now, that's this, delete the racy photo!, and tell a caring and responsible adult about it, like me! You can control your finger from pushing the send button to forward the photo. But if you don't and you choose to send it or it's accidently sent and you're caught, painful experience has taught you that there's a really good chance you'll get felony charges and be wearing a sex offender label. Ben grinned and said. "Not happening!"

A Father's Sexting Teen

Blog Entry 30 of 30 – Everybody's Talking It's a Big Deal
Tuesday | March 10, 2009 | 4:45 p.m.

Photographers from "People" Magazine were in our living room late yesterday afternoon snapping pictures of Ben and his buddy John who went on "Inside Edition" with us. The sexting article will be out the end of this month. Can't believe that this whole thing has gone has far as it has. The more articles I read, I see how much sexting is still going on at school campuses across the country. It's at high schools, junior highs and at elementary schools. I just read online about the research that Susan Lipkins, a psychologist who specializes in bullying. She says that kids as young as nine years old may be sexting. Now that's scary Don't let anyone tell you that sexting is not a BIG DEAL. IT'S A BIG DEAL! Some teens and preteens know how dumb it is to go there but there are those who don't have a clue and say "it's fun," and just "a joke." They think the way Ben used to because they have a blindfold on about the ugly fallout from sexting. Few realize, including parents of teens, that most states have outdated laws that will be used to unfairly prosecute a teen sexter by using child pornography laws to arrest and charge the teen with producing, distributing and possessing pornographic material of a minor. If the charges stick, then the guilty teen is put on a Registered Sex Offenders list and he'll be on it for up to forty years! Here's my rant one last time! Child pornography laws were put in place to punish adult predators from sexually harming and taking advantage of children. Sexting teens should not be in the same "prison cell," as adult sex offenders who are found guilty of preying on and committing foul and indecent acts with children. If teens are caught sexting they should be fined and charged with a misdemeanor and made to do community service and only in extreme cases where there is evidence of predatory and malicious intent, should they be arrested, or kicked out of school.

    Lawmakers in several states are debating penalties. The Utah Legislature passed lighter penalties making sexting not a felony but a misdemeanor. Several states are trying to change the laws so that teens who sext are not put in the same category as adult child pornography offenders. Parents, start an honest conversation with your kids about sexting, what it is and what can happen if they sext. Ask them this question. Before you send a text with a photo attached would you want the whole school including the Principal, Vice Principal, fellow students in your math class, or your parents, grandparents, brother, sister, police, the county attorney, or anyone who goes online to see that text message? If not, then don't send it because once your teens "special" picture that they innocently think will only be seen by that very wonderful someone and it leaves their cell phone, or computer there is NO getting it back, ever! Make sure your teen knows that sending out a photo into the cyber world, is like losing a limb, once its gone, its GONE!

Finally, don't be afraid to tell your teens very simply: Sexting is dangerous and dumb. To sext is to be "sexually active" and teens doing it are not in control. When teens sext, they are sharing intimate and private pictures of their bodies with another person and they have no control over who will see them or where their photo might end up. Let your teen understand that they or someone they know could be sitting at a ball game and their racy picture pops up on the stadium billboard and it's seen by "everyone." Your teen's scandalous photo could end up on a "kiddie" porn site and ogled at by pedophiles and peeping tom types who don't need binoculars to better their view. They can blow up your teen's picture as largeas they want on their computer screen.

Please don't be one of those parents who proudly say "my teen would never sext, because they have way too much sense, or self respect to ever do that." I read about one mother who said that her 15-year-old daughter would never sext because she had way too much respect for herself. The truth was that this same girl had already sent a naked video of herself to her boyfriend almost two years earlier and it was already on a bunch of amateur pornography sites across the internet. The mother was clueless about her daughter living with the shame of her secret. Be aware of what's going on with your teen, their cell phones, social networking sites, email accounts or any digital device that can send or receive messages.

Take the time to really get to know your teen. Read online somewhere that when a teen family member feels connected to their parent or guardian because the parent is talking the teens "love language," (what makes the teen feel loved and cared for) then the chance of a teen going into crazy and stupid behavior is not as high. Enjoy family dinners together! Learn to play and have pure fun! Get "wild and crazy" have a water balloon or a pillow fight. Give your teen the shock of their life and crank up their music and listen to it with them, and for the really daring parent you might want to try a few of their dance moves! Don't forget you were a teen once!

You won't regret living those moments! Time to get off my soapbox, Kim is calling. She probably wants me to run to the store before I start cooking dinner. Until my limo business picks up again, I manage the house and take care of family issues, while Kim works during the week at the library. My sweetheart is the best! She's been a very patient and understanding wife throughout this ordeal. I chose a keeper, for sure.

One more thing, after dinner tonight, I'll do my best to go out and shoot some hoops with Ben, (my pain from the surgery is not as bad). This I know without one doubt! I'm glad that I'm a father of a former sexting teen.

P.S. On March 30, 2009 "People" Magazine published a story about my former sexting teen, Ben and his friend John. By the way, John and the other four students got off the hook, and I've heard that they're all former sexting teens too!

EPILOGUE

It's been 10 months 24 days since the final August 27, 2009 letter from the Court arrived completely releasing Ben and the other students from the charges. The Clerk Magistrate wrote:

August 27, 2009 Dear Mr. Benjamin Hunt,

Please be advised that you have successfully completed the conditions imposed on you with regards to the above referenced docket numbers and you have not since violated the law. I am pleased to inform you that you will not be prosecuted for the offenses that brought you into this court. Since your case did not go forward for prosecution, you will not have a criminal record. Please accept my best wishes for your future.

Very truly yours,
Charles P. Andrade,
Jr. Clerk Magistrate

After the letter arrived, a few interesting things happened worth writing down. Ben's cell phone was finally returned. The girl's picture was still on it! I quickly deleted it! An officer from the Falmouth Police Department was fired for sexting. I guess the problem of sexting can be much more serious for an adult. I'm not a big guy anymore. I lost one hundred and ten pounds. I had gastric bypass surgery and the doctor discovered that when he went in I had an adhesion to a prior hernia repair which was not sewn up properly. It was that which caused the jabbing pains in my side after my diverticulitis surgery. My doctor told me that if he had not fixed it, I would have likely been dead in thirty days, because toxins from my colon were leaking into my system. The old saying, "he's just full of it," was really true for me. Ben and I are featured in a Human Relations Media DVD about sexting. The DVD is being distributed in High Schools and Jr. High Schools across the country. Ben and I also went on Dr. Phil's show. I was happy that he thinks it's a "really, really good idea" to put a warning message in cell phones for kids and teens that explains what might happen to them if they're caught sending a racy photo. Kim still works at the library and she's still the love of my life. We'll be celebrating our fifteenth wedding anniversary soon. Since she liked the Statue of Liberty snow globe so much, I think I'll buy her a Brooklyn Bridge one. I probably ought to get her a new diamond for her wedding ring; otherwise I might get "clobbered" with the globe.

Limo business hasn't picked up too much. I'm staying busy working around the house on lots of different projects including Kim's "honey do" list!

Today I'm focused on spreading the word to teens about the bad fallout that comes with sexting and bullying. It's been my goal to educate parents and create awareness about the dangers associated with it. I'm happy to be working with Annie and her publisher to build momentum towards any national grassroots effort to stop sexting and bullying. Check out **NowStopSexting.org** and **NowStopBullying.org** for the latest information, tips and news on these important subjects.

The websites will also offer downloads of Annie and her friend's Keep It On rap song and No Vultures poster. Teens can hang up the posters around their school so students know a Bully Free Zone is going on! Teens With Dreams posters and resources will also be available in the near future to encourage teens to live their dreams and build their character right.

I'm staring a whole lot less at my popcorn ceiling before I fall asleep. I'm trying to make it a habit to talk with God every night and even during the day. I've discovered that I'm not worrying as much now. I'm seriously thinking about taking the family out on Sunday and going to church more often.

I can't believe I did this. I dusted off my Bible and began to read a little bit and one of the first verses I read was, "The fear of the Lord is the beginning of wisdom." Because of what I've been through and what I may go through in the future, I don't think I can ever get enough wisdom. I certainly need it to raise a teen right, to keep 'em on the straight and narrow so they don't go too far south and get their "toast burned."

Life Building Insight

What parents can do to prevent teen sexting

Contact your teen's cell phone provider and ask that a block be put on the phone that will prevent pictures from being sent or received either as a text or email attachment. Most cell phone companies have this feature. It's very important to check with your cell phone company about other parental controls they offer to help keep your teen safe. The best parental control you can have with your teen is to do whatever you can to nurture a positive relationship, one that is built on mutual trust, respect and understanding.

Sit down and talk with your teen about responsible cell phone use. Tell them having a cell phone is a privilege not a right and help them understand both the benefits and dangers of owning a cell phone and that they should never abuse it by sending or receiving inappropriate photos via email and text messages. Tell your teen that if they abuse it they will lose it! Make sure you follow through and do what you say otherwise your teen won't respect you. Draw up a cell phone contract. Clearly define your rules and expectations so that your teen understands what you expect of him or her. Check out **www.aFathersSextingTeen.com** for a free download of a teen cell phone contract.

Discuss the following ten risks with your teen and then reread Blog 26 "I wish I hadn't" teen guys and gals talk about their experience with sexting.

Teens who "sext" run the risk of one or more of the following happening to them:

1) Being caught, arrested and charged with child pornography. Federal law prohibits the distribution of nude or semi-nude photos of minors, even if the minor receives a nude photo (doesn't create or send it) and forwards it to another minor.

2) Registering as a sex offender.

3) Experiencing shame and embarrassment should their racy photos circulate among peers and fellow students.

4) Having their racy photos go viral and attracting the attention of stalkers and sexual predators.

5) Becoming the object of cruel bullying both at school and online, where they could be called a "ho," a "slut," or far worse names.

6) Expulsion from school and other extracurricular activities, sports, cheerleading, running for school office.

7) Humiliating and hurting your loved ones who may receive your racy photos on their cell phones should the boyfriend or girlfriend breakup with you and then decide to be mean and revengeful towards you.

8) Losing friendships because your friend's parents may not want you around their teen son or daughter as they may now think of you as a bad influence on their child.

9) Feeling great despair and devastation over your photo

being out and then getting bullied so you tragically consider ending your life.

10) Losing the respect and acceptance of a future employer or college admission director who may do a background check on you and discover your racy photos floating around in cyberspace.

Teach your teen to be Web Wise Check out **www.WebWiseKids. org**. A cutting edge group which launched the award winning **Missing** program reaching millions of middle-school kids across the country. **Web wise** offers on-line high tech computer game simulations based on real-life criminal cases to help your teen protect themselves from the ruses and traps of on-line predators.

Really listen to your teen and not to just the words they're saying but to what's behind them and what they're not saying. Studies show that parents are the last to know if their child has been bullied. You can make sure this doesn't happen to your teen by listening and initiating conversation every day. Try not to come across like a courtroom interrogator. Drilling your teen with a bunch of questions and then not listening with your heart to them feels disrespectful and like your only on a fishing expedition for information not relationship. Try doing something fun with them, starting a special project together that sincerely interests your teen. Spend time getting to know them, understanding their likes, and dislikes from the music they listen to, to their favorite TV shows and movies. Always come alongside as a caring parent and be real and transparent.

Let them know you are there for them and that no matter what you love them!

Be your teens best cheerleader Find creative ways to build them up and be specific in your praise. If you've hurt them by anything you've done or not done, be humble enough to say the "I'm sorry" words. Be quick to praise and slow to criticize. Be specific with whatever good words you have to say about their efforts at school or anything they've done that reflects hard work and discipline. Don't just say, "Good Job" but tell your teen what was good about the job he or she did. For example: "I like the shade of blue that you used on your poster. It's a nice contrast with the yellow. I've noticed how much effort you've put into this project and I also like how you didn't put off your work. I am very proud of you."

Eat family dinner together and keep the **fun** in it! Make mealtime fun family time as a top priority. Check out a great website that offers free family conversation fun ideas. **www.MakeMeal-timeFamilyTime.com** It's a wonderful tool to help keep your family dinner conversation lively and interesting. There's nothing like a thought provoking question or idea to stir up a good discussion and interaction with those at your table. Teens and younger children develop a sense of security and belonging when they make contributions at mealtime and enjoy connection with other family members. Research proves that families who frequently eat meals together have children who are more healthy and well adjusted than those families who don't eat together. The family that eats together, plays together, serves together, **stays** together. On serving together, volunteer as a family to work at a soup kitchen. Serve the poor and needy it will do your heart well and you'll discover that you and your family's heart will grow ten sizes. I know of a family who goes to Mexico three times a year to serve battered women and I've see how close, fun and caring this family is.

Hug, laugh, and build each other up with kind words. It's really true what the proverb says. "Good words are healthful to the body and healing to the bones." Turn off the technology (computers, cell phones and TV) at least one day a week if you can! Try the Amish lifestyle even if you don't have a horse and buggy! Remember what the Amish know, less is more, simple is good and real is right. Be fun as a family. Roast marshmallows over the fireplace, make s'mores, play Monopoly,® Scrabble,® Scattergories® or any other fun game. Try charading each other, cooking, bowling, or doing anything else as a family that is **Pure Fun!**

Go to church as a family and encourage your teen to attend a youth group. You'll be surprised that the Bible's wisdom works for real life and if followed will give you a successful life.

Keep your teen busy you don't want them in a place of "nothing to do." Keep them focused in working towards goals. Get them involved with activities from sports, to drama or whatever they enjoy doing that's positive. Help your teen **dream** and discover his/ her passion.

Be a positive role model for your teen. Be transparent about who you are and be willing to admit mistakes when you make them. Remember, your teen's eyes are watching you, seeing how you act and talk with your friends, family and acquaintances. The old saying, "the apple doesn't fall far from the tree" is really true. Remember, more is "caught" than is "taught." Guard your character and teach your teen the power of a good choice and the negative fallout from a bad one.

Post the following on your refrigerator. It's a great reminder for everyone in your family:

Watch your **thoughts**, for they become your words.

Watch your **words**, for they become your actions.

Watch your **actions**, for they become your habits.

Watch your **habits**, for they become your character.

Watch your **character**, for it becomes your **destiny**.

   — **Anonymous**

Life Building Insight

What parents can do if your teen is caught sexting

Assure your teen that you love them and want the best for them even if their sexting behavior displeases and embarrasses you. Discipline your teen in love.

If your teen is facing child pornography charges don't give up the fight. Consult with an attorney in your area. The current laws on child pornography and sex offenders are not written for sexting teens. Many states are changing the law and making teen sexting a misdemeanor charge.

Make an appointment to see a good family counselor. Contact a local church for a referral.

Life Building Insight

Eight things to know if you're a sexting teen

1) Know that a Provocative Picture (PP) will ignite lust in the one who sends it and in the one who views it. Lust is not love and it will never be your friend. Research reveals the danger of lust out of control. If lust is consistently fed by looking at PP(s), watching on-line pornography, or engaging in non-marital sexual relations, then you are on a slippery slope of compromise which is a fire that could consume and destroy your life. Some of the negative fallout might be unwanted pregnancies, a broken heart, abuse, abandonment from the one who was supposed to love you, sexually transmitted diseases, loss of self respect and the respect of those who know and truly care about you.

2) Know that your PP(s) will likely be distributed over cell phones and social networks across the world, downloaded by pedophiles and show up on pornographic websites and you can never rewind or delete your PP(s).

3) Know that you will likely become a victim of vicious gossips and mean-spirited bullies who make it their pleasure to launch Hell's flaming arrows both in cyberspace and the real world after your PP(s), are viewed by them. Then they will begin a 'trash you' campaign starting with cruel jokes, insensitive innuendos, snide remarks, and crass conversation. Such heated and painful attack will cause you severe emotional trauma and your pain will be so great that you might contemplate taking a rope and hanging it around your neck (**Note**: No matter how horrible you feel **NEVER** make this choice. There is hope, healing and deliverance. Talk to a trusted counselor, friend, parent or relative.)

4) Know that the significant someone who asked you to send a PP(s) will dump you. A disappearing act he/she will become.

5) Know that anyone who asks you to send a PP(s) of yourself is not your true love or even your friend. They are a "Frenemy," one who appears to be a friend but is really an enemy because they are selfishly using you and treating you like an object for their lustful pleasure.

6) Know that you are breaking child pornography laws by sexting and if you are found out you could be charged and prosecuted as a sex offender pornographer and struggle with the real possibility that you may serve jail time for producing, distributing and possessing PP(s) of an underage minor, be it of yourself or another. Having been prosecuted and found guilty you will be put on the Registered Sexual Offenders list which will be on your record for at least the next ten years or more. A child pornography conviction carries weightier penalties than most hands-on sexual offenses. The Federal Adam Walsh Child Protection Act of 2007 requires that sex offenders as young as fourteen be registered.

7) Know that if you continue on the path of sexting by allowing yourself and others to be aroused by PP(s) thinking that it is "harmless" and not hurting anyone that you are deceiving yourself. Your character is on the path of becoming vulgar and indecent and you won't be someone your future children or grandchildren can be proud of.

8) Know that choosing to be **pure**, respecting and honoring yourself and the one you care about is a **wise** choice that will reap lifetime rewards. Don't believe the lie that everybody is doing it! Tons of teens all over the world are committing to purity. They understand an amazing and incredible secret, that **Purity is Power!** Being pure in thought, mind and body will allow one to be powerful, wise and strong.

Real Voices For Pure Fun Powerful Life Thoughts

Aaron Craver

Former NFL fullback

Youth Life Coach

"When Annie first asked me if I would be one of her **Real Voices for Pure Fun**, I didn't know quite what she wanted. But after talking with her and being impressed that she was trying to make a difference in teens lives for good by encouraging them to stay away from behavior that's destructive and promotes a "deathstyle" not a "lifestyle,"

I said sure, how could I not since I also have a passion to see teens build their lives on a solid foundation of truth, going for their dreams, developing their talents, and doing what's right.

I grew up in Compton and life wasn't easy. I struggled with asthma but I always tried my best in everything I did. I learned early that if I worked hard both in school and out, developed my athletic ability, did my best to have **Pure Fun,** and not "mess myself up fun," and had faith in the One who made me, that things would work out, one way or another.

Strange as it was, I never thought of having a career in football. I played baseball in high school and it wasn't until a friend convinced me to give football a try that I did. I transferred from a small community college to Fresno State and in two years I rushed for 2,251 yards and 26 TDs. Later while on the Denver Broncos I was John Elway's "go to guy" on the short third down passes. My ten-year career as an NFL player taught me that the effectiveness of my success on the field of football was in direct correlation to how well

I practiced, learned the plays, and kept the rules of the game with a positive attitude. My success in the game was always linked to the level of my commitment to be prepared and ready for action before the game. The more prepared I was the better I played. This is true for life. As a sports trainer and community leader, I tell teens that life is like a big football field. You have to know your position, play it well, know your teammates, work together, know the rules, follow them and be ready for action, because the surprise tackle may come along in the next play. Be on your guard!

I also make it one of my goals when I speak to teens to encourage them to take their life seriously by not wasting their time but to spend it wisely developing their unique God-given talents and skills. If teens do this, they'll eventually discover that the greatest life is the life that's given away by helping others, striving for excellence, working hard, following the rules and enjoying one's life by having **Pure Fun! Pure Fun** is about those activities and pursuits that keeps a teen in his game yet still provides a healthy relaxing and refreshing time out.

My message to teens is to keep away from the "bad stuff" which may give you a fast "feel good" but in the end you'll be tackled and taken down. In other words, you'll be what every quarterback in the NFL doesn't want to be, **SACKED!**

Annie Winston

Real Voices For Pure Fun Powerful Life Thoughts

Kandee Johnson

Makeup Artist | Style Seeker |
Heart Inspirer

Teen guys and teen gals,

Save yourself for marriage.

Sometimes we are seeking to be **so** loved, to feel **so** desired, **so** approved and **so** wanted and we find ourselves standing on that doorway of having sex, and in one moment of giving in you can forever change the course of your life and a child's life.

We've now become **so dumb!** By having sex outside of

marriage, you don't win the guys affection or respect, you lose that guard over your heart, you lose the control over your emotions. You've given a gift to someone who should have committed before God, family and friends to be with you 'til "death do us part" and "to have and to hold you forever" This someone in the heat of fiery passion or in the moments leading up said to you, "yes, yes, yes, baby, I love you so much, you mean everything in the world to me," and then months later those words ring like an empty drum, hollow, and you've discovered you played the fool, your "forever guy" now wants to dump you. You also discover that he has found another "lover" to believe the lie he told you, "yes, yes, yes, baby, I love you so much, you mean everything in the world to me." Now, you're alone and left with all this horrible emotional baggage to sort out. I say most sincerely; **protect yourself! Be wise! Guard your heart, guard your love, and guard your decisions!** No matter how hard it gets or how exciting the moment of being with your guy is, **guard your decisions and guard your body!** Put it under lock and key until God sends your "Knight in Shining Armor." This guy will "rock in his love for you" he will win your heart, mind and devotion because he shows you his good character and you can see it over time and he proves just how wonderful, trustworthy and respectable he is towards you. Don't give your precious pearl of purity, your gift, to a "pig," a selfish using guy. Your gift is for your wedding night to the man of your dreams, an unselfish caring guy. Having sex is like icing on the cake, but if there's no cake (a marriage between two committed people) then eventually you'll wake up and learn that "nothing is there," just like the Emperor who woke up to the fact he wasn't wearing any clothes but not before he stupidly denied the obvious.

I iced the cake at seventeen, knew it wasn't the right way, but did it anyway thinking that he "loved me so much" and that somehow our love would work out, but it didn't. I allowed my wild emotions for a guy to take over my brain and then my life. I thought I was so in love. I gave up a lot of things because I got all "love crazy" over him. I now know that real love should never make you feel "love crazy," it should just add to your already beautiful life. My "love-crazy" emotions caused me to give up things like going after my career in modeling and my college education. My "love crazy" feelings caused me to rush to make it right between my guy. We got married and I moved away with my new eighteen-year old husband. I remember my agent at my modeling agency telling me "Kandee, if you do this you'll be throwing your modeling career away!" He was right. As I sat, seventeen-years-old, in my hot, dark, tiny apartment in Phoenix, with no money or friends, and a "husband" who said he couldn't live without me, but now only wanted to be with his friends, I realized...I made a **HUGE** mistake. If I had just dated him and said, "you know what, you don't get the gift, I'm saving it for my husband. No 'bon-bon' from the Kandee store until we're married."

I probably would have blown him out but I also would

have blown out his game which would have been a good thing! If I had made this choice, my heart would have never had to hurt as bad as it did when we broke up. I've learned my lesson, **true lasting love waits!** That saying about food, "No food tastes as good as it feels to fit into your skinny jeans," well, "No sex feels as good as it feels to have control over your heart, your life and your emotions." Say **NO** to **SEX** outside of marriage and build an awesome future for yourself and while your building, don't forget to care about others, enjoy your life and always try your best to live in a way that's just **Pure Fun!"**

Real Voices For Pure Fun Powerful Life Thoughts
Jim Burns

President of Homeword | National Radio Host | Author

"I've worked with teens and families for decades. I know this: teens want truth, what's real and they can't stand hypocrisy either in themselves or in someone else.

Live a positive transparent God honoring life and help your teen understand the importance of saving themselves for marriage and to not treat their body like a garbage disposal allowing anyone who comes along with declarations of love and devotion, rights to it. Encourage them to honor

The Purity Code

In honor of God, my family, and my future spouse, I commit my life to sexual purity. This involves

Honoring God with **my body**.

- Renewing **my mind** for good.
- Turning **my eyes** from worthless things.
- Guarding **my heart** above all else.

All studies show that the more positive, value centered sex education kids receive, the less promiscuous they will become. Help your teen understand that they will have pressures from what they watch on television, listen to in their music, and see at the movies, to be sexually involved with someone they "like" and are in a casual or committed relationship with. Their emotional involvement with someone they care about often exceeds their ability to make wise and healthy decisions about how they should act. Parents, help out your teen! Teens learn best when they talk. Give them your ears and listen. Let them know that the best life is the one that honors God and honoring God is to be pure not messing yourself up with having sex before marriage. Being pure does not mean you have to be a big bore! You can be fun because you know your limits and you're committed to not crossing them! You have a network of friends and trusted family to keep you accountable. A smart wise teen is all about having **Pure Fun!** They understand that short term "happy hits" are not worth the high price of sexually transmitted diseases, dealing with unwanted pregnancies or wearing the label of sex offender. Teens have **Pure Fun!** Parents help your teens keep their **Fun Pure!**"

Real Voices For Pure Fun Powerful Life Thoughts
Randy Largent

CEO of EFX – Because It Works | Youth Sports Coach

"Kids and teens are our future. I love being a husband and a father of a teen. I care about teens and the issues they face because life is not easy, especially when they're starting out.

I'm committed to encourage youth to have the winning edge in everything they do and to have the best shot at a life that's wise, healthy and productive. Coaching sports is one of the ways I do this.

Sports is a fantastic way to teach valuable life lessons to kids and teens. Beginning with the importance of playing by the rules, valuing the whole team's effort to get the win, going for it and striving for your personal best. I've coached lacrosse, football, baseball and basketball and have put more balls, bats, helmets into kid's hands than I ever thought I would. I've given my best effort to teach the necessary skills and mind set to my players so that they're a winner in every game be it on a grassy field or the concrete of life. I tell them in lots of different ways to "Never give up! A **winner** has a winning attitude and a winning attitude always wins whether the scoreboard says so or not!"

I've taken my passion for helping youth grow by teaching positive attitudes and true athleticism in sports to my work as CEO of EFX. My company distributes and markets the highest quality performance technology. It is amazing to see how Olympic athletes and ordinary kids and teens are given the "winning edge" of energy, balance and focus by wearing our wrist bands. As a parent, coach and businessman, I've learned the necessity of doing your best to keep your child free from being thrown off balance by drugs, alcohol and sexting. Parents do all you can to protect your most incredible and priceless asset, your children! Encourage and build them up with kind, good words. Make sure your discipline is done out of love not anger.

Have **Pure Fun** and **enjoy** life together. You'll never live to regret that decision! Such positive memories won't be taken away no matter what happens, even if the stock market crashes or your home forecloses. What's truly valuable and important in life has very little to do with material success power or position. Never forget that, because loving on and having **Pure Fun** with your wife, kids and teens is way more important!

Real Voices For Pure Fun Powerful Life Thoughts

Annie Winston

Author | Pure Fun Lover | Proud Mom of Three Teens

"I love fun, especially **Pure Fun**. I'm one of those types that can almost never have enough good clean wholesome fun! I love to laugh and get a bit "crazy" with my three teens. No matter how "rich" or "poor" we are or have been, we'd always try our best to laugh and have fun together. I remember a time when our electricity had been turned off, and it wasn't going to get turned back on until the next day. So that night, we lit candles, sat around and played charades. I think it was one of our best times together. I remember laughing so hard that I almost peed my pants. Another time, a pipe in the kitchen burst and a good part of our house was flooded. Huge blowers were set up to dry us out but most of the carpet had to be removed. For the next several months we lived on cement floors, (couldn't afford to put in new carpets) but we had "fun" because we pretended that we were "camping" and we'd laugh about it, making the best of the "disaster." No matter what happens, I encourage my kids to be thankful. I tell them, "A thankful heart is a happy heart."Trying times will always come one's way but an "attitude of gratitude" makes them almost disappear. I'd always make a special effort to have super fun birthdays for my kids from the time they were little to when they were teens. It was important for me to tell them just how special each of them were and how happy and excited I was that they were born! I didn't have a whole lot of money to spend, so their birthday celebration had to be creatively low budget. I remember my oldest daughter's ten-year-old birthday party. Her birthday was in December and when we sat around planning it on a cool fall day in early November, she told me she wanted a beach party with kayaking. I said, "Honey but your birthday is in December and I don't know if we'll get a sunny beach day." She insisted that was what she wanted so I said okay! I told her we should ask God to give us a very sunny day for your birthday. We prayed. Her birthday arrived. It was one of the hottest days (90 plus degree) on record for the month of December. God abundantly answered her prayer! Her birthday was **Pure Fun** and full of sun! Her swim coach let us borrow his kayak so all her friends went out kayaking. There's nothing better than loving your kids and making sure that they're having **Pure Fun!** Make it a priority to have bunches of it! Doing so never goes out of style and best of all you won't forget **Pure Fun** moments. You'll treasure them as some of the choicest memories of your life! \

Sponsored Links

www.KandeeTheMakeupArtist.blogspot.com

www.homeword.com www.efxusa.com

www.AnnieWinston.com

www.MakeMealtimeFamilyTime.com

www.WebWiseKids.org

www.NowStopSexting.org

www.NowStopBullying.org

DOWNTOWN CAMPUS LRC

J.S. Reynolds Community College
3 7219 001615981

HQ 799.2 .C45 W56 2010
Winston, Annie.
A father's sexting teen

CPSIA information can be obtained at www.ICGtesting.com
224841LV00002B/94/P

9 781456 334444